**Jörg Matthias Determann** is Assistant Professor of History at Virginia Commonwealth University in Qatar. He is the author of *Historiography in Saudi Arabia: Globalization and the State in the Middle East* (I.B.Tauris, 2014) and a joint winner of the 2013 BRISMES Leigh Douglas Memorial Prize.

# RESEARCHING BIOLOGY AND EVOLUTION IN THE GULF STATES

Networks of Science in the Middle East

Jörg Matthias Determann

I.B. TAURIS

LONDON · NEW YORK

Published in 2015 by
I.B.Tauris & Co. Ltd
London • New York
www.ibtauris.com

Library of Modern Middle East Studies 167

ISBN: 978 1 78453 156 0
eISBN: 978 0 85772 944 6

A full CIP record for this book is available from the British Library
A full CIP record is available from the Library of Congress

Library of Congress Catalog Card Number: available

Typeset in Garamond Three by OKS Prepress Services, Chennai, India
Printed and bound by CPI Group (UK) Ltd, Croydon, CR0 4YY

*To my family, friends, colleagues and students*

# CONTENTS

# LIST OF FIGURES

# ABBREVIATIONS

| | |
|---|---|
| Adach | Abu Dhabi Authority for Culture and Heritage |
| ADCO | Abu Dhabi Company for Onshore Oil Operations |
| ADMA-OPCO | Abu Dhabi Marine Operating Company |
| Aramco | Arabian American Oil Company |
| BGI | Beijing Genomics Institute |
| BRGM | Bureau de recherches géologiques et minières |
| ECWP | Emirates Center for Wildlife Propagation |
| ENHG | Emirates Natural History Group |
| GCC | Gulf Cooperation Council |
| IFHC | International Fund for Houbara Conservation |
| IUCN | International Union for Conservation of Nature |
| KACST | King Abdulaziz City for Science and Technology (formerly SANCST) |
| KAU | King Abdulaziz University |
| KAUST | King Abdullah University of Science and Technology |
| KKWRC | King Khalid Wildlife Research Center |
| KSU | King Saud University |
| MAW | Ministry of Agriculture and Water, Saudi Arabia |
| NARC | National Avian Research Center |
| NCWCD | National Commission for Wildlife Conservation and Development |
| NHM | Natural History Museum |

| | |
|---|---|
| NSF | National Science Foundation |
| NWRC | National Wildlife Research Center |
| SANCST | Saudi Arabian National Center for Science and Technology (later KACST) |
| Socal | Standard Oil of California |
| SQU | Sultan Qaboos University |
| TAVO | Tübinger Atlas des Vorderen Orients |
| UAE | United Arab Emirates |
| UNESCO | United Nations Educational, Scientific and Cultural Organization |
| USDA | United States Department of Agriculture |
| WWF | World Wildlife Fund |
| ZSL | Zoological Society of London |

# ACKNOWLEDGEMENTS

In order to complete this book on scientific connections in the Gulf, I benefitted from a dense network. When I studied Arabic at the University of Vienna, Rüdiger Lohlker provided some early inspiration for my research through a course on Islamic creationism in 2006. In addition, Herbert Eisenstein served as a model through his research on Arabic zoography. I started planning this project as a graduate student at the School of Oriental and African Studies (SOAS), University of London, in 2010. During this time, I was fortunate to have been a visiting scholar at the King Faisal Center for Research and Islamic Studies in Riyadh. This centre represents an island of academic efficiency in the Gulf – like some of the biological research centres described in this book. Like the National Wildlife Research Center and the King Khalid Wildlife Research Center, it also enjoys the patronage of the children of King Faisal. I am very grateful for the quick and comprehensive support provided by the centre's staff, especially Yahya Ibn Junaid, Ibrahim Al-Hadlaq and Awadh al-Badi.

From Britain and Saudi Arabia, this project accompanied me to Germany and Qatar. Between 2012 and 2013, I completed an early draft of the manuscript as a postdoctoral fellow at Zentrum Moderner Orient and the Berlin Graduate School Muslim Cultures and Societies, Freie Universität Berlin. I am extremely grateful to Ulrike Freitag and Gudrun Krämer for this fellowship. Regarding my time in Berlin, I am also very grateful to Veronika Lipphardt for inviting me to the meetings of her research group at the Max Planck Institute for the History of Science. I undertook subsequent revisions of the manuscript while serving as a faculty member at Virginia Commonwealth University in

Qatar. This transnational institution, and others within Education City, represent further islands of efficiency within rentier academia. I am very grateful for the academic freedom and support provided by Qatar Foundation. I also thank Allyson Vanstone, Aisha al-Muftah, Byrad Yyelland, Dina Bangdel, Leah Long, Line Christiansen, Melis Hafez, Radha Dalal, Sadia Mir and my other colleagues and friends in Qatar for their support. Lore Guilmartin kindly ordered numerous books for my research for the VCUQatar Library. Special thanks go to my students for their inspiring discussions about the history and sociology of science in the Muslim world.

Outside of Qatar, numerous people provided me with feedback at various stages of my research, helping to make the manuscript richer, stronger and more accurate. I am extremely grateful to the anonymous peer-reviewers commissioned by I.B.Tauris for their helpful comments. I also benefitted from the great expertise and eye for detail of my editor Maria Marsh and her colleagues at I.B.Tauris. Before the formal peer-review, I was very fortunate to receive constructive feedback from various teachers and colleagues. Marwa Elshakry, Jonathan Ercanbrack, Mike Farquhar, Ben Fortna, Konrad Hirschler, Martyn Smith, Melania Savino and Philipp Wirtz all read the first proposals for this project in 2012. Husnul Amin, Jeanine Dağyeli, Eva-Maria Silies and Cyrus Schayegh, as well as anonymous reviewers commissioned by Deutsche Forschungsgemeinschaft commented on the subsequent proposals in 2013. A number of my friends and colleagues, including Luigi Achilli, João Rangel de Almeida, Elise Burton, Birgit Krawietz and Dan Stolz, kindly commented on different parts of the manuscript. Furthermore, I had the privilege of discussing a chapter with students and colleagues at the Berlin Graduate School Muslim Cultures and Societies in 2013. In the final stages of preparing the manuscript, a number of biologists and protagonists of this book commented on different chapters. They include Zaher Al-Sulaimani, Faysal Bibi, Andreas Buerkert, Shaukat Chaudhary, Yolanda van Heezik, Shahina Ghazanfar, Harald Kürschner, Friedhelm Krupp, Stéphane Ostrowski, Phil Seddon, Sevket Sen and Joe Williams. Oliver Jarvis of Global Proofreading and Copy-editing kindly aligned the manuscript with the I.B.Tauris house style and corrected my English.

In addition, various audiences have provided me with very helpful feedback on the presentations of my research. In 2012, I introduced outlines of the project at seminars at the Berlin Graduate School Muslim Cultures and Societies and Zentrum Moderner Orient. In 2013, I presented the initial results at Virginia Commonwealth University in Qatar. Furthermore, I had the pleasure of discussing my research at various workshops and conferences. I benefitted from the comments and questions of the interdisciplinary audience at the second IBZ Symposium at Internationales Begegnungszentrum der Wissenschaft, Berlin, in 2013. I am also very grateful to Paul Aarts and my audience for their comments after a presentation at the workshop 'In Search of the Kingdom – Emerging Scholarship on Saudi Arabia' in Berlin during the same year. Finally, I presented my research at the Qatar Foundation Annual Research conference in Doha in 2013. By analysing the challenges and possibilities of rentier science, I hope that this book contributes to the knowledge economy that Qatar Foundation and other institutions in the Gulf seek to create.

I also thank my interviewees for their time and for the large quantity of unpublished material from their personal archives. Many of my interviewees kindly sent me resumes of their lives and careers. Faysal Bibi provided me with copies of newspaper clippings and rare documentaries about paleontological work in Abu Dhabi. Harald Kürschner and Joe Williams sent me photographs from their work in Saudi Arabia. I also thank Daisy Cunynghame and Kate Tyte for their help in accessing the archives of the Natural History Museum in London. Furthermore, I am grateful to the Institut für Geschichte der Arabisch-Islamischen Wissenschaften in Frankfurt am Main, Germany, for hosting me in June 2014. I especially thank Gesine Yildiz and Norbert Löchter for helping me access the institute's rich collections, and Fuat Sezgin, Mazen Amawi and Farid Benfeghoul for their inspiring conversations.

Finally, I thank my parents, Sibylle and Michael, my brothers, Christian and Claudius, and the members of my extended family for all their love, interest and support. Regina, Johannes and Simon Kienecker kindly hosted me during my research trips to Tübingen and helped me access the rich library collections of the University of Tübingen. My grandfather, Fritz Determann, a mathematician with broad interests in nature, has long been a source of scientific inspiration.

# TRANSLITERATION, TRANSLATION AND REFERENCES

In writing Arabic names, I have tried to use forms that readers can recognise and look up easily in encyclopaedias and on the internet. When a common Romanisation or translation of the names of Arab individuals and institutions exists, I follow it. When Arab authors have a preferred way of spelling their names in Latin characters, for instance, in their publications in European languages, I follow this preference. For common Arabic words that have entered the English language (such as ulema), I use the forms in the *Oxford Dictionary of English*.[1]

For the bibliographical details, I follow the Romanisation standard of the American Library Association and the Library of Congress (ALA-LC).[2] I hope that this will allow readers to find the literature easily in library catalogues. I use the same standard for the transliteration for Arabic words that have not entered the English language and for Arabic place names without a common Romanised form. However, in the main text, I avoid using diacritics for proper names, as is practised by the *International Journal of Middle East Studies*. All translations, unless otherwise acknowledged, are my own. Unless declared otherwise, I have been the interviewer in the interviews and the recipient of emails cited in the footnotes. All amounts given in dollars are in US dollars. Citations broadly follow the notes and bibliography format in the *Chicago Manual of Style*.[3]

# DRAMATIS PERSONAE

Abdul Mon'im Talhouk – entomologist at the American University of Beirut and the Saudi Ministry of Agriculture

Abdulaziz Abuzinada – Saudi botanist at King Saud University, president of the Saudi Biological Society, and secretary general of the National Commission for Wildlife Conservation and Development (NCWCD)

Abdullah bin Abdulaziz – Saudi crown prince and, from 2005, king

Abdul-Rahman Khoja – Saudi botanist at the National Wildlife Research Center (NWRC)

Ahmed Boug – Saudi primatologist at the NWRC

Andreas Buerkert – German botanist at the University of Kassel

Fahd bin Abdulaziz – Saudi crown prince and, between 1982 and 2005, king

Friedhelm Krupp – German zoologist, editor of *Fauna of Saudi Arabia*

Ghazi Sultan – Saudi deputy minister of mineral resources

Harald Kürschner – German botanist involved in Tübinger Atlas des Vorderen Orients (TAVO)

Jacques Flamand – French scientist at the NCWCD

Jacques Renaud – French veterinarian, founder of Reneco and consultant manager of the NWRC

Joe Williams – American zoologist at Ohio State University and the NWRC

Kamal Batanouny – Egyptian botanist at KAU and Qatar University

Michael Gallagher – British ornithologist and advisor at the Oman Natural History Museum

Qaboos – sultan of Oman

Ralph Daly – British advisor to Sultan Qaboos of Oman

Saud Al-Faisal – foreign minister of Saudi Arabia and managing director of the NCWCD

Shaukat Chaudhary – Pakistani botanist at the Saudi Ministry of Agriculture and Water

Sultan bin Abdulaziz – chair of the NCWCD

Wolfgang Frey – German botanist involved in TAVO

Yolanda van Heezik – ornithologist at the NWRC

Zayed bin Sultan – ruler of Abu Dhabi and president of the United Arab Emirates

# CHAPTER 1

# SCIENTIFIC GULF

In March 2001, Saudi Arabia's supreme religious authority, the Permanent Committee for Research and Legal Opinion, issued a fatwa banning the popular Japanese game Pokémon. The fatwa said that besides being a means of gambling, 'the game contradicted Islamic teachings by promoting Darwin's theory of evolution'.[1] The scholars referred to the growth of weaker pocket monsters into stronger ones, a process that is called 'evolution' in the game. Following the fatwa, a senior official in the Saudi Ministry of Commerce declared that Pokémon products would be removed from the shelves and destroyed. Saudi customs officials would receive instructions not to allow Pokémon products into the country.[2] Within weeks, other Gulf states followed the Saudi ban. In April 2001, the Qatar-based 'global mufti'[3] Yusuf al-Qaradawi issued a similar fatwa against Pokémon in order 'to preserve our children, their faith, their morale and their money'.[4] The Islamic Affairs and Charitable Activities Department in Dubai also took action against the Japanese game. During the same month, the department issued a fatwa declaring Pokémon as having a negative influence on children. The Dubai fatwa also found that the game 'promoted Darwin's theory of evolution which is rejected by Islam'.[5]

These bans in the Gulf were part of global attempts to control the Pokémon craze. In America, many schools banned trading in Pokémon cards because it distracted pupils from their studies.[6] However, in the Gulf the banning of Pokémon was also part of wider rejections of evolution by ulema. In one of its fatwas, the Saudi Permanent Committee for Research and Legal Opinion under Abdulaziz ibn Baz declared that 'the theory of evolution known as Darwin's theory contradicts the Qur'an, the Sunna and the consensus (*ijmāʿ*) of scholars.'[7] In a subsequent fatwa, the committee declared it 'false' that 'man had once been an ape and evolved', citing a Qur'anic description of Adam's creation as evidence.[8] In 2011, Yusuf al-Qaradawi maintained a similar position in his show *Sharia and Life* on Qatar's Al Jazeera television network. He asserted that 'man was created directly and did not evolve out of any other species'. Al-Qaradawi added that 'this is consistent with the Qur'anic theory and with religious beliefs in general, even Christian and Jewish religious beliefs'.[9]

Given the authority of scholars, such as the Saudi grand mufti Abdulaziz ibn Baz, it is not surprising that creationism was present in science textbooks published by the Ministry of Education and in the National Museum of Saudi Arabia. A sixth-grade general science textbook from 2007, for instance, quoted Qur'anic verses about God's creation of man. A twelfth-grade biology textbook from 2006 stated that 'in the West appeared what is called "the theory of evolution" which was derived by the Englishman Charles Darwin, who denied Allah's creation of humanity, saying that all living things and humans are from a single origin'. The textbook added, 'We do not need to pursue such a theory because we have in the Book of Allah the final say regarding the origin of life, that all living things are Allah's creation.'[10] When I visited the National Museum in Riyadh in 2010, it displayed fossils from different periods in the Earth's history, but did not mention evolution. Instead, a text stated that 'each plant and animal has been created by Allah with a defined purpose and place in the web of life on the planet'. The museum text continued, 'At different times in the Earth's history the same sort of ecological place in the web has been filled by different animals and plants.'

The challenges by ulema and textbooks have made evolution a sensitive or 'touchy'[11] issue in the Gulf. In higher education, some professors taught the theory,[12] but not without controversy. In 2007, students at the College of Arts in Dammam, Saudi Arabia, complained about the appearance of 'Darwin's theory' in a textbook on human geography. The students were especially concerned about 'claims that humans originate in apes'. An administrator promptly defended the college, stating that the textbook had been 'chosen by the professor after the agreement by the department's council'. In addition, the students were not 'required to believe in what appears in the book and to adopt it'. The administrator added that the head of department would 'alert the students' to the controversial points in the book and 'make the right view clear to them'.[13]

What is surprising, however, is that an environment in which evolution was so sensitive still provided careers for proponents of evolution. One of these evolutionists was the Saudi biologist Abdulaziz Abuzinada. Born in 1940, he was one of the first professional biologists in the Gulf monarchies. Common among the pioneers of his generation, his biography reads like a series of firsts. After his graduation in 1965, he became one of the first graduate assistants in the natural sciences at the recently established King Saud University (KSU) in Riyadh.[14] In 1971, Abuzinada completed a doctoral thesis in botany at the University of Durham in the United Kingdom. Thereafter, he became the first Saudi chair of KSU's Department of Botany and one of the first Saudi full professors in the natural sciences. Although trained as a botanist, he became a leader of the entire biological profession in Saudi Arabia. In 1974, Abuzinada co-founded the Saudi Biological Society and became its first president. In 1986, he also became the first secretary-general of the Saudi National Commission for Wildlife Conservation and Development (NCWCD). Despite the sensitivity of evolution, Abuzinada did not appear to have suffered in his career, when, in 1998 and 2003, he began two of his articles with the following statement: 'The Earth's genes, species, and ecosystems are the product of over 3000 million years of evolution, and are the basis for the survival for our species.'[15]

How was it possible for biologists in the Gulf, such as Abdulaziz Abuzinada, to employ evolution despite contrary views by senior ulema? As evolution forms a central theory of the modern life sciences – alongside cell theory and genetics – this is an important question for the wider study of science in Arabia. How and to what extent scientists in the Gulf were able to work with one of the most important and controversial scientific theories arguably has implications for the Gulf monarchies' potential to innovate and create a knowledge economy.[16] This book thus examines the history of modern biology in general, and evolution in particular in the Arab Gulf monarchies. I give special emphasis to the employment of evolution and related concepts such as natural selection, adaptation, competition and survival in the publications of biologists who worked in Arabia. I also ask what enabled scientists to employ evolution in the political, religious and social, as well as natural environments of the Gulf. In answering this latter question, I give special attention to the role of networks between scientists and other actors, such as princes, plants and animals. These connections have a history of hundreds of years.

Modern biologists from the Gulf region, such as Abdulaziz Abuzinada, continued a longer history of Arabic-speaking scholars of animals and plants. Scholars in the medieval Arabic-lands, like some of their modern counterparts, often enjoyed patronage from princes interested in falcons, horses and other animals. Moreover, pre-modern as well as modern Arabic-speaking scholars adopted knowledge from foreign sources, especially in European languages such as Greek and English. A few medieval scholars also had a recognisable 'scientific approach' and undertook 'experiments'.[17] Al-Jahiz, a scholar born in Basra in the eighth century, even offered accounts of changes in animals that bear resemblance to the theory of evolution.[18] Although Linnaean taxonomy largely replaced pre-modern classification systems from the eighteenth century onwards, many older Arabic names for animals survived, and some even entered European languages. The English name for the main prey in falconry, the houbara bustard, for instance, derives from ḥubārá, its common name in medieval Arabic literature.[19]

As modern biologists in the Gulf often received parts of their education in Europe and America, they also drew on previous Western explorations of Arabia. One of the first explorations that classified Arabian organisms according to Linnaean nomenclature was an expedition that included Carsten Niebuhr and Peter Forsskål, a Swedish student of Carl Linnaeus. Forsskål spent six weeks in Jeddah in 1762 describing Red Sea fish, before sailing south to Yemen. Christian Gottfried Ehrenberg and Wilhelm Hemprich, naturalists from the University of Berlin, continued Forsskål's work, travelling to Jeddah and the Asir mountains between 1824 and 1825.[20] During the same period, the Gulf became the 'Arabian frontier'[21] of British interests in India. As early as 1787, the East India Company requested from its agents in the Gulf specimens of six plant species, including date palms, for a new botanic garden in Kolkata.[22] During the time of the British Raj, other naturalists and political agents sent specimens from the Gulf to the empire's botanic gardens and published their findings in periodicals such as the *Records of the Botanical Survey of India* and the *Journal of the Bombay*

*Natural History Society.*[23] These publications in Western languages formed reference works for subsequent generations of biologists, including Arabs and Muslims.

Much of the research undertaken by Europeans in Arabia between the mid-eighteenth and the mid-twentieth century served national and imperial aims. Expeditions sought to provide European states with prestige and knowledge about non-European regions, including information on their strategic and economic potential. Natural knowledge and specimens acquired from outside Europe also allowed institutions such as the Kew Royal Botanic Gardens to become imperial 'centres of calculation'[24] for natural information. Finally, the description and classification of plants and animals according to the Linnaean system allowed European elites to feel 'in command' of foreign environments and natural resources.[25]

Alongside national aims, Western explorations of Arabia had transnational dimensions. Although Forsskål's expedition was sponsored by the Danish king for nationalist reasons, it included Germans and – with Forsskål himself – a Swede. It sought to answer a catalogue of questions about Arabia compiled by a privy counsellor to the king of Britain and prince-elector of Brunswick-Lüneburg.[26] Furthermore, the expedition depended on an ideology of scientific transnationalism or cosmopolitanism. It thus already featured some of the national and transnational aspects of modern 'cross-boundary science'.[27] Many naturalists themselves believed that they were not just serving their respective governments or nations, but humanity as a whole. Influenced by Alexander von Humboldt's expedition to South America, Ehrenberg wrote that 'the idea of a voyage is my ideal'. He added, 'It is my firm belief that all good for mankind must derive from natural history.'[28]

Cosmopolitan ideals continued in the form of international development and collaboration that shaped much of Western research in Arabia in the decades of decolonisation after World War II. During the 1950s, the Arabian American Oil Company (Aramco) brought entomologists to Saudi Arabia in order to control malaria as part of its development programme.[29] In order to develop their economies, the Gulf governments hired other Western companies to survey the natural resources in their territories, including plants and animals. These surveys enabled the Gulf states to expand their oil industries and agriculture, as well as their control over nature and society.[30] At the same time, the scientists involved in these surveys contributed to the global knowledge about Arabian flora and fauna. Italian botanists, for instance, described the flora of Saudi Arabia as part of agricultural development surveys during the 1960s and 1970s. These surveys were commissioned by the Saudi government and carried out under the United Nations Development Programme. The botanists published their results in an Italian journal, but in English, making it accessible to a transnational community of scientists.[31]

Scientists also relied on, and incorporated, local knowledge that had developed over hundreds of years. Such ethnoecological knowledge thus formed another background for modern biology in the Gulf. Bedouin lore formed one of the bases – together with Greek literature – of pre-modern Arabic books on animals.[32] Forsskål, who had studied Arabic at Göttingen, and his successors frequently Latinised local Arabic names in classifying species. In an article for the *Journal of the Bombay Natural History Society*, Violet Dickson, the wife of a British political agent in Kuwait, added to a list of Kuwaiti wild flower descriptions according to their use by Bedouin.[33] Aramco's Arabian Research Unit hired part-time Bedouin 'relators' (*ruwāt*, singular *rāwī*), using an old Arabic word different from 'informant' to avoid intelligence-gathering connotations. The knowledge of these relators allowed Jim Mandaville, an Aramco Arabist, to publish books entitled *Flora of Eastern Saudi Arabia* and *Bedouin Ethnobotany*.[34] Professional biologists who had grown up in the Gulf, especially in rural areas, often retained some of this ethnoecological knowledge. This knowledge did not contain an elaborate theory of evolution as such. However, intimate knowledge of the survival of plants and animals in the desert perhaps prepared some Gulf scientists to accept that organisms 'adapted' to their environments.

In the decades after World War II, indigenous professional biologists joined expatriate scientists and development experts in Arabia. In order to develop their countries and decrease reliance on foreigners, the Arab Gulf monarchies established universities, beginning in the 1950s. One year after its foundation in 1957, King Saud University comprised departments of botany and zoology. Kuwait University offered degree programmes in botany and zoology from its establishment in 1966.[35] Qatar University and Sultan Qaboos University in Oman also taught the biological sciences after their opening in 1973 and 1986,[36] respectively. Although most of the initial professors at these universities came from outside the Gulf, they quickly graduated a first generation of native professional biologists. These professionals often completed their studies with postgraduate degrees abroad. Between 1964 and 2005, the Saudis alone completed 200 doctoral dissertations in agricultural, veterinary, botanical, zoological and other biological sciences at American universities.[37]

The teaching of modern biology at the university level familiarised a significant number of Gulf nationals with the theory of evolution for the first time. While Arabs had read Charles Darwin in various ways in the Levant and Egypt between 1860 and 1950,[38] his reception in the Gulf was very limited. In largely illiterate societies, *al-Muqtataf*, the periodical that had first introduced evolution to Arabic readers, had only a few readers and contributors in the Arabian peninsula.[39] Texts by the Muslim reformers Jamal al-Din al-Afghani and Muhammad Abduh, who sought to refute Darwin,[40] were read in Mecca and perhaps Riyadh, but had little impact during the first half of the twentieth century.[41] From the 1950s onwards, however, at least the Gulf nationals studying biology at universities in the West would have learned evolution as a central theory.

While the establishment of an indigenous biological profession turned some people from the Gulf, such as Abdulaziz Abuzinada, into evolutionists, evolution became a sensitive issue during the second half of the twentieth century. In 1964, a Saudi scholar published a book entitled *Islam and Darwin's Theory*, which sought to encourage Muslim youth to defend Islam against the materialist aspects of the theory.[42] Around the same time, the ideology of the Muslim Brotherhood spread in the Gulf, including the hostility of its leading member Sayyid Qutb to evolution. During the Arab Cold War between Egypt and Saudi Arabia, the Saudi government gave refuge and positions to members of the Muslim Brotherhood who were persecuted in Egypt. The Brotherhood thus became very influential in Saudi education,[43] and the teachings of Darwin – alongside Karl Marx and Sigmund Freud – became severely restricted.[44] The hostility of much of the education system towards Darwin's theory resulted in a low acceptance of evolution among Saudis. A poll by Ipsos from 2011 found that 75 per cent of Saudis were creationists who believed in God's creation of human beings. 7 per cent were evolutionists who believed that humans had evolved from 'lower species such as apes'.[45] In my experience, probably more people in the Gulf believed in evolution than were willing to admit in polls or public. However, many would consider human evolution a sensitive topic.

Because of the sensitive nature of evolution, scholars of Islamic intellectual history have paid more attention to it than to most other modern scientific theories, including relativity.[46] Many of these scholars have analysed the writings by popularisers and critics of evolution, especially journalists, editors, ulema and writers.[47] Other scholars have examined the teaching of evolution through textbooks, curricula and surveys.[48] Thanks to all these researchers, much literature exists on the popularisation, translation and critique of the theory of evolution in the Eastern Mediterranean in particular. The Arab Gulf countries have, however, so far been rather neglected, as they have in studies on intellectual history and the history of science in general.

Moreover, while a considerable amount of literature on Darwin's reception by non-scientists exists, we have little scholarship on the employment of evolution by professional biologists in the Middle East. This focus on non-scientists is also true beyond the field of biology. Many scholars of modern Middle Eastern history have investigated the attitudes of ulema, writers, journalists and intellectuals towards science and modernity.[49] Other scholars have shown that, far from being apolitical and value-neutral, science has served as a means of state control and social transformation.[50] However, while this research has given us a better understanding of science in the Middle East, it has been state-centred. It has also tended to describe the states in the region as rather monolithic actors that are separate, and often in opposition, to society. In contrast, this book seeks to provide a better understanding of the works and lives of Middle Eastern scientists and their manifold connections with the state and society.[51]

By examining the history of modern science and scientists in the Gulf, this book also seeks to go beyond the wholesale and often negative views of scientific production in the contemporary Arab world. Ahmad Dallal, provost of the American University of Beirut, has described the basis for scientific research in contemporary Muslim societies as 'extremely impoverished'. He added that 'in the absence of a living scientific culture in the modern Muslim world, we can discuss discourses on science, not science itself'. In 2003, the British magazine *New Scientist* reported that because of either religion or despotism 'science in Islamic countries has all but ground to a halt'.[52] Narratives of a golden age and decline in Arab and Muslim science arguably shaped such views. They recounted that Arab and Islamic civilisation enjoyed a golden age during the Abbasid caliphate in Baghdad. During the early modern period, the rise of the West occurred simultaneously with a decline of the rest, including Islamic civilisation. For at least several decades, Middle Eastern and world historians have criticised these narratives of a golden age and decline.[53] Yet, this criticism has not led to the complete disappearance of views that science 'declined' or 'stagnated' in the Arab and Muslim world at some point during the second millennium.[54]

While this book seeks to avoid the tropes of decline and stagnation, it also cautions against too celebratory and uncritical views of recent scientific projects in the Gulf. These views also referred to the narrative of a previous golden age in Arab and Muslim science. In 2009, King Abdullah University of Science and Technology (KAUST) opened on the Red Sea coast. In his opening speech, King Abdullah stated that the university 'did not emerge from nowhere. It is a continuation of what distinguished our civilization in its golden age.'[55] Around the same time, Qatar Foundation expanded its Education City, which hosted branch campuses of major American, French and British universities. These universities included Weill Cornell Medical College and Carnegie Mellon University, both of which have taught biology in Qatar. An article about Education City in the British newspaper *The Telegraph* from 2012 was entitled 'A new golden age rises under the desert sun.'[56]

A detailed analysis of the growth of modern biology in the Gulf provides a better foundation for assessing the prospects of science in the Arab world than the narratives of a golden age and decline. Some scholars have already investigated the political context of the establishment of KAUST and Qatar's Education City. Toby Jones interpreted KAUST as a continuation of the Saudi state's long-standing attempt to control nature and society through science and technology.[57] Christopher Davidson considered the Western branch campuses in Qatar and Abu Dhabi as a means for Gulf monarchies to gain 'soft power'.[58] However, in order to understand the actual research undertaken at Gulf universities, it is important to go beyond the perspectives of the states and to study the scientists. These scientists also formed part of a scientific culture, but one that was arguably more transnational than essentially Muslim or Arab.

## Arabian science

Through their extraordinary wealth, the Gulf countries form an important case for studying science in the contemporary Arab world. With some of the world's largest oil and gas fields at their disposal, the Gulf monarchies were able to invest more resources in science and higher education than other Arab countries. The cost of constructing Qatar's Education City totalled $33 billion,[59] while KAUST possessed an endowment of $10 billion.[60] Financial resources alone do not determine scientific productivity, and some Saudi institutions received the label 'expensive white-elephant universities'.[61] Yet, the Gulf monarchies also possessed access to other resources. As they remained allies of America and the United Kingdom after their independence, they imported a lot of Western technology and expertise. With the exception of the Gulf War, wars and revolutions in the Middle East also affected the monarchies to a lesser extent than the Arab republics. Because of all these factors, the Gulf monarchies acquired the technological, human and financial resources to surpass older Arab centres of learning, such as Egypt, in the more expensive fields of science. Decades before the establishment of Qatar's Education City and KAUST, the Gulf attracted scientists from Western as well Arab countries. As early as 1980, Ziauddin Sardar noted in the journal *Nature* that the 'Middle East brain drain switches back from the West'.[62]

The Gulf monarchies also make a good case study for studying biology and evolution as a flashpoint in the relationship between science and religion. Conservative Islam in general had an 'uneasy relationship' with modern science.[63] One conservative Islamic current, the Wahhabi mission, became very strong in Saudi Arabia and Qatar from the eighteenth century onwards. The Wahhabi-dominated Permanent Committee for Research and Legal Opinion in Saudi Arabia was particularly conservative regarding social norms. Although the committee was generally open to modern technology, especially in medicine,[64] its head, Abdulaziz ibn Baz, was notorious for asserting that the Earth was flat and that the sun moved around the Earth.[65] When the geneticist Theodosius Dobzhansky wrote that 'nothing in biology makes sense except in the light of evolution', he cited Ibn Baz's attempt to suppress Copernican theory, calling the sheikh 'ignorant' and 'hopelessly biased'.[66]

Despite Ibn Baz's notoriety, Arabian governments were not hostile to science per se. The Gulf perhaps became 'sectarian',[67] but it was also scientific. For much of Islamic history, religious and natural knowledge were hard to separate.[68] The motto of Kuwait University and King Fahd University of Petroleum and Minerals was 'Say: Lord, increase my knowledge.' The Arabic word for knowledge in this Qur'anic verse is *'ilm*, which encompasses religious as well as natural knowledge. Gulf rulers often supported scholars who argued for the compatibility of Islam and science. In the 1970s, King Faisal of Saudi Arabia appointed the French Muslim Maurice Bucaille as his family physician. In 1976, Bucaille published a book entitled *The Bible, the Koran and Science*, in which he argued that the Qur'an, unlike the Bible, does not contradict scientific facts.[69] His arguments echoed those of the earlier Egyptian scholar Muhammad Abduh, who had argued that Islam, in contrast to Christianity, was not in conflict with science.[70]

Despite these views on harmony between science and Islam, the Gulf states did not escape the influential notion that 'modern science' was the equivalent of 'Western science',[71] and had materialist aspects. This led the Gulf monarchies to support an intensifying debate about the Islamisation of science or the creation of a 'specifically Islamic form of science'[72] during an era of decolonisation. Saudi Arabia became a forum for this debate by hosting institutions for intra-Islamic dialogue, including the Muslim World League and the Organisation of Islamic Cooperation. Saudi officials were particularly willing to embrace an Islamised science to bolster their Islamic credentials in competition with other Muslim-majority countries. During the Arab Cold War with Egypt, the Saudi support for an Islamic form of science also took place in the context of a wider struggle against atheism, socialism, materialism and revolutionary forms of Arab nationalism.

Helped by its proximity to the Muslim World League and the Organisation of Islamic Cooperation, King Abdulaziz University (KAU), with branches in Jeddah and Mecca, adopted the Islamisation of science. During the late 1970s, the university tried to create an 'Islamically-oriented conception of knowledge'. KAU's aims included 'the replacement of Western knowledge with an authentic Islamic-oriented knowledge', and 'the production of individuals who are Muslims in spirit and practice'. In order to achieve these aims, KAU, like other higher education institutions in the kingdom, introduced a number of compulsory core courses in 'Islamic culture' for students of all disciplines.[73] An important figure behind KAU's policies was Abdullah

Nasseef, who served as secretary general and director of KAU before becoming secretary general of the Muslim World League in the early 1980s. Offering an example of Islamised science, Nasseef published a chapter with a 'theogeological approach'. He first quoted the French mystic Simone Weil, stating that 'a science which does not bring us nearer to God is worthless'. Nasseef then cited verses from the Qur'an as evidence that God produces colourations in rocks by sending down rain. The geologist concluded his chapter by arguing that 'scientists, particularly Moslems and Arabs, would benefit from reading the holy Qur'an by paying due attention to the verses that point to God's power and the "how" he created things in origin'.[74]

Nasseef's 'theogeological approach' was similar to 'scientific exegesis', another school that sought to combine Islam and science. This school emerged in the nineteenth century with ulema such as the Egyptian Muhammad Abduh, who tried to prove the compatibility of the Qur'an with modern science. During the twentieth century, successors of Abduh even claimed to have discovered references to the Big Bang Theory and modern embryology in the Qur'an. The Permanent Committee for Research and Legal Opinion in Saudi Arabia rejected such interpretations of the Qur'an based on empirical science, fearing the subordination of the Qur'an to scientific theories.[75] Nevertheless, scientific exegesis became popular in the Gulf. In 1984, the Muslim World League established a Commission on Scientific Signs in the Qur'an and Sunna in Mecca. In 2006, Mohammed bin Rashid, the ruler of Dubai, awarded the Islamic Personality of the Year Award to Zaghloul El-Naggar, a prominent Egyptian scientific exegete.[76] Six years earlier, the same award had gone to the Qatar-based mufti Yusuf al-Qaradawi.[77]

Although the Gulf governments supported science and its combination with Islam, restrictions on freedom of expression hindered this trend. In 2007, the journal *Nature* complained that Saudi Arabia had 'one of the world's worst track records on academic freedom'.[78] At the United Arab Emirates University, temporary, renewable contracts for many faculty members, combined with 'administrative oppression', resulted in self-censorship and the avoidance of 'sensitive issues'.[79] In Kuwait, pressure from administrators and self-censorship made classroom discussions that impinge 'even tangentially on religion a potential mine field'.[80] Some professors in Saudi Arabia believed that conservative religious and government informers monitored their classroom comments.[81] Criticism of this situation arose inside the Gulf monarchies themselves. In 1980, a Saudi newspaper reported on a symposium at King Saud University on the duties and rights of Saudi professors. It quoted a Saudi geologist, saying that it is difficult to engage in productive research when 'there is no freedom of opinion or freedom of thought or freedom of anything'.[82]

Despite complaints by some professors, pubic academic institutions often participated in censorship. As experts on specific subjects, some of their members reviewed publications for the government. In libraries, committees dealt with censorship issues and responded to objections raised by users about materials.[83] In Saudi Arabia, the King Abdulaziz City for Science and Technology (KACST), which served as the country's science academy and national laboratory, became responsible for filtering internet traffic in 1997.[84] Saleh Al-Athel, the president of KACST, announced that his organisation would 'prevent the entry of material that corrupts or that harms our Muslim values, tradition and culture'.[85]

Although academic freedom was restricted, censorship did not affect all areas of academic inquiry in the Gulf. That Saudi newspapers reported complaints by professors, for instance, suggests that the censorship had not been entirely suffocating. While paper copies of foreign books and journals, like other imports, needed to pass customs, electronic versions did not. Moreover, scientific and technical materials were often exempt from censorship. Even if certain books were controversial, many libraries still acquired them, making them accessible in separate rooms.[86] Despite the sensitivity of evolution, works on the theory were available on library shelves at King Saud University, Kuwait University and Qatar University, among others. Internet censorship mainly targeted pornographic, oppositional, anti-Islamic and human rights websites, affecting scientific content to a lesser extent.

Apart from a general prohibition of criticism of the ruling families, censorship was, however, not uniform across the Gulf countries and at all times. Kuwait, for instance, enjoyed more press freedom than its neighbours did between the 1960s and 1986, when the government established restrictions on political dissent. In 1992, after the Gulf War, the government lifted censorship again, but with some restrictions remaining. The United Arab Emirates also offered journalists a more open environment than other Arab countries, leading the Saudi-owned news channel *Al Arabiya* to establish itself in Dubai rather than Riyadh.[87] The branch campuses of foreign universities in Qatar's Education City also enjoyed more freedom than many national universities in the region.

While recognising differences between and within countries, it still makes sense to study the Gulf monarchies as a group. They were all members of the Arab League and ruled by powerful and long-standing dynasties.[88] As elsewhere in the Middle East, Britain and the Ottoman Empire had shaped the creation of the modern states and their borders during the late nineteenth and early twentieth centuries.[89] The British Empire also exerted varying degrees of control over the Gulf's coastal regions until 1971. However, members of the ruling families had often lived in the Gulf for hundreds of years. In the eighteenth century, the House of Saud founded the first Saudi state in central Arabia, and the houses of Sabah, Khalifa and Said gained power in Kuwait, Bahrain and Oman. During the nineteenth century, the houses of Nahyan and Maktoum established themselves in Abu Dhabi and Dubai. In Qatar, British support allowed the House of Thani to gain pre-eminence over other emirs and sheikhs only in the twentieth century. Nevertheless, members of the Al Thani had reached the rank of sheikh even before the British started interfering in Qatari politics.[90] This was in contrast to the royal families of Jordan and Iraq, who only arrived in these countries with British support during the early twentieth century. Contemporary support for science by members of long-standing dynasties thus also appears as a continuation of a tradition of princely patronage for scholarship in the region.

Adding to their shared monarchical history, the Gulf states witnessed the establishment of large oil and gas industries during the twentieth century. In 1932, drilling crews from Standard Oil of California (Socal) discovered oil in Bahrain, stimulating the search for, and eventual discovery of, oil among its neighbours. Socal, together with Texaco, subsequently established Aramco. The rents that these companies paid to monarchs reinforced existing patterns of distribution based on British payments, creating 'rentier states'[91] and 'societies of intermediaries'. The ruling families distributed the oil revenues among their subjects through long chains of brokers, whose interests were often individualised, informal and local instead of class-based.[92]

Oil money also gave ruling families the resources to buy off or suppress opposition and to respond to societal demands without sharing power. However, while considerable research exists on the political systems produced by oil and rents,[93] we know much less about the sociology of science in the Gulf. It has been assumed that the reliance on oil rents produced a 'rentier mentality', in which financial reward was more related to patronage and personal connections than to hard work and risk-taking.[94] (One could argue, however, that the maintenance of social connections itself also required emotional and other forms of labour.) Oil money also allowed for the import of expertise and technology, often on a turnkey basis, without creating the necessity to develop local capabilities.[95] Rentier culture has thus been seen as undermining research and innovation.[96] These negative views ignore, however, examples of scientific success and innovation in rentier states in the Gulf and elsewhere, such as Norway. While recognising the constraints posed by rentier culture, this book investigates the forms of science that allocation states enabled and gave rise to.

Historically well-established in local societies and with many resources at their disposal, the Gulf's ruling families were very resilient. They survived a wave of Arab revolutions, which toppled the monarchies of Egypt, Iraq, Libya and Yemen during the 1950s and 1960s. In 2011, they collectively survived another wave of Arab uprisings that brought down the heads of state of Tunisia and, again, Egypt and Libya. Despite rivalries and occasional conflicts, the Gulf monarchies helped each other survive at critical moments. During the Gulf War, Saudi Arabia, Qatar and the United Arab Emirates formed part of the coalition forces that drove Iraqi forces out of Kuwait. In 2011, forces from Saudi Arabia, Kuwait and the UAE helped put down protests in Bahrain. For science, regime stability meant institutional continuity and expansion. Rather than suffering from political disruptions, scientific institutions benefitted from the persecution, purges and wars that drove people from elsewhere in the Arab world to the Gulf.

Based on their shared culture and political systems, the Gulf monarchies also engaged in significant collaboration in fields other than security, such as science and the environment. In 1975, a conference of Gulf ministers of education established an Arab Bureau of Education for the Gulf States in Riyadh. This bureau in turn founded a common institution of higher education, the Arabian Gulf University, in Bahrain in 1981. The name of the university and the choice of Bahrain were an expression of solidarity directed against previous Iranian claims over the

Bahraini archipelago.[97] In 1983, the Arab Bureau of Education for the Gulf States also launched the *Arab Gulf Journal of Scientific Research*. Its founding associate editor was the Saudi evolutionist Abdulaziz Abuzinada. Furthermore, the Gulf monarchies committed themselves to cooperation in wildlife protection. In 1982, Bahrain, Kuwait, Oman, Qatar, Saudi Arabia and the United Arab Emirates established the Cooperation Council for the Arab States of the Gulf, also known as the Gulf Cooperation Council (GCC). In 2001, its member states signed a Convention on the Conservation of Wildlife and their Natural Habitats, in which they agreed to cooperate in 'research, exchange of expertise and training of specialist personnel'.[98]

## Science in transaction

Despite the uneasy relationship between modern science and conservative Islamic currents, it is not surprising that modern biology and evolution as one of its central theories entered the Middle East, including the Gulf. Darwin's ideas successfully penetrated most regions of the world. In places as diverse as Kolkata, Tokyo and Beijing, scholars reconciled these ideas with local traditions, although not without some struggle.[99] Scholars have long recognised the global reach of modern science. As early as 1967, *Science* magazine published an article that proposed a three-stage model for 'the introduction of modern science into any non-European nation'.[100] Scholars have since criticised this particular model as Eurocentric, and proposed transnational and global histories of science instead. Rather than following the spread of science from the West to the rest, such transnational histories have put emphasis on networks of exchange and circulation.[101] Through connections between pre-modern Arabic zoological knowledge, ethnoecology, and Western colonial science and development expertise, the growth of modern biology in the Gulf was also a transnational story.

Not just modern biology, but also the sensitivity of evolution had a global history. For many adherents of Abrahamic religions, man's common ancestry with apes was difficult to reconcile with the notion of man as a creation distinct from animals and angels.[102] In the Middle East, perhaps the earliest controversy over evolution arose through a transnational missionary connection with American Christians. In 1882, Edwin Lewis, a Harvard-educated professor and missionary, cited

Darwin as an example of scientific achievement in a commencement speech at the Syrian Protestant College in Beirut. This led the college's American president and board of trustees to force Lewis to resign.[103] More recently, influential Muslim creationists, such as Harun Yahya in Turkey, also used transnational connections. During the 1980s, Yahya started challenging evolution in pamphlets, often recycling tropes of American Christian creationists and proponents of intelligent design. Subsequently, he became one of the most visible and best-funded representatives of Islamic creationism. In the 1990s, Yahya established the Science Research Foundation that propagated his ideas in books and web pages.[104]

Informed by transnational perspectives on the history of science, this book examines the networks that have enabled and shaped modern biological research in Arabia, including research in the sensitive area of evolution. This book thus also draws on a growing literature on networks. During the 1960s, sociologists and social psychologists started paying close attention to social networks.[105] Twenty years later, Bruno Latour and other proponents of the actor-network theory widened the study of networks to comprise non-human and technical actors.[106] While Latour's analyses have been mainly qualitative, other scholars have since also sought to quantify scientific networks by using citation indexes and databases.[107] Working qualitatively or quantitatively, sociologists have provided important information on the centrality of certain authors and institutions as centres of calculation or hubs. This book seeks to combine the network approach used by sociologists of science with research on the sociology and political economy of the Gulf states. I rely on conceptions of rentier states by scholars, such as Hazem Beblawi and Matthew Gray.[108] I also draw on Steffen Hertog's work on the hub-and-spoke system of Saudi politics and the importance of intermediaries in chains of patronage.[109] I am thus especially interested in how networks created scientific 'islands of efficiency'.[110]

In addition to research in sociology and political economy, this book draws on scholarship on networks in the history of science. Like the sociologists, and often influenced by them, historians have increasingly emphasised connections and webs in world history.[111] As part of this trend, historians have conceived the British Empire as being a multimodal network rather than a structure consisting of a single centre and a periphery.[112] Based on the work by these imperial historians, as well as sociologists of science such as Bruno Latour, historians of science have given increasing attention to networks.[113] A focus on networks has strengthened the view of the history of science as a socially contextual history rather than a history of ideas.[114] Historians of scientific networks have also contributed to decentring and globalising our view of modern science. Instead of focusing on European scientists alone, they have shown the connections between scientists in Europe and their collaborators in the Americas, South Asia, China and Australia. However, they have paid hardly any attention to scientists' networks in the Middle East. As a result, the Middle East still appears peripheral to histories of modern science.[115] This book seeks to rectify this view.

The concept of networks has also attracted criticism and should be used with caution. To the British historian Simon Potter, for instance, networks suggest a rather loose structure. Such a structure was not necessarily the case in the British empire, which had 'a tendency towards systematization'.[116] In particular, the connections between London and each of the colonies were often stronger than the connections between the colonies themselves. Potter's criticism does not prevent the use of the concept of networks. However, it reminds us that within a network not all connections are of equal strength. This means that connections between some institutions in the Gulf and institutions in the West, such as between a branch and a home campus, were stronger than between individual institutions in the Gulf.

How do networks explain the history of biology and evolution in the Gulf? The education of scientists in Arabia depended on networks, particularly between domestic and foreign universities. Gulf universities recruited most of their initial faculty members and administrators through networks with Arab and Western countries. Many of these professionals came from Egypt, where the establishment of the Egyptian University (later Cairo University) in 1908 had preceded that of King Saud University, the first university in the Gulf monarchies, by half a century. Partly thanks to the oil wealth, the newly established Gulf universities were able to attract faculty members from countries with this advantage in modern education. Between 1977 and 1987, Egyptians made up an average of around 47 per cent of the PhD-holding faculty at the United Arab Emirates University. The next most populous group comprised the Iraqis, at a steadily declining average of around 16 per cent.[117]

These networks with Egypt and other countries also provided avenues for Gulf nationals to gain an education abroad. In the 1950s and 1960s, many people from the Gulf studied in Cairo as one of the educational and scientific hubs of the Arab world. The UAE's first minister of education was himself a graduate of Cairo University.[118] With increasing oil revenues and strong political connections, America too became a more affordable and attractive destination for Gulf Arabs. At that time, the United States increased its intake of foreign students and, partly because of immigrating scientists, assumed leadership in global science. Even traditionally British-oriented countries, such as Australia, sent more students to America in the decades after World War II.[119] In order to facilitate entry into American graduate programmes for Gulf nationals, universities in the region imitated American rather than Egyptian educational models.[120]

Having completed their education abroad, Gulf biologists often continued to use and expand their networks with foreign institutions. Some of these institutions, such as Kew Gardens, had themselves long been central nodes of imperial and global scientific networks.[121] They had collections and libraries that surpassed those of any young research institution in the Gulf. They were thus able to provide numerous services to Gulf-based scientists, such as identifying specimens, supplying literature, commenting on draft papers and helping with publication. These networks also served as conduits for information about modern biology, including evolution.

The Gulf's oil wealth helped its young institutions establish and sustain networks with much older institutions in the West, exchanging money for expertise. 'Science in action'[122] in the Gulf was thus also science in transaction. Oil money from the Gulf helped Western institutions maintain their leading positions. In 1980, for instance, Princeton University agreed to assist King Saud University in building a centre for graduate and postdoctoral training and research in the life sciences. In return, the Saudi university donated $5 million to Princeton. Princeton used this money, perhaps the biggest single grant from an oil-producing state to an American university until that time, to modernise its laboratories and support faculty in biology.[123] Other payments were much more modest, but also aimed at incentivising cooperation. In 1984, for instance, Kuwait University paid a botanist at London's Natural History Museum (NHM) around £80 for refereeing a grant proposal on 'The Marine Algae Flora of the Kuwait Coast' prepared by one of its faculty members.[124] The Gulf's scientific institutions frequently also provided visiting collaborators with 'first-class' hotel accommodation and transportation.[125]

Science in transaction does not mean that Gulf monarchies merely purchased and imported Western science. Instead, human and non-human actors in Western as well as Gulf countries co-produced science. Middle Eastern oil was crucial to the wealth and prosperity of the West and the funding for science. After the oil boom of 1973, the Gulf states were no longer peripheral, but central nodes in the global flow of energy, money and people.[126] Science in the Gulf was thus neither Western nor Arab and Islamic, whether in terms of 'locality' or 'essence'.[127] Instead, science in transaction was transnational, with scientists, specimens, instruments and publications all crossing national borders. Even if the biggest funders of science were national agencies, such as the Qatar National Research Fund, the scientists themselves were often cosmopolitan and their methods globalised.

Oil was not the only natural resource that fuelled and sustained the Gulf's scientific networks. For research on biology, animals, plants and their environments were just as valuable as money. London's Natural History Museum, for instance, informed the agricultural department of the Kuwait Oil Company that it would 'be pleased to accept any specimens from the area of Kuwait for identification'. In 1950, the company thus forwarded duplicates of its collection to the museum, hoping that they would be 'of some value'. The company also asked the keeper of the museum's botany department to recommend a textbook for the identification of plants. In 1977, Michael Gallagher, a British advisor to the Omani government, requested access to the NHM's skin and wing collection for the illustration of a book on the birds of the sultanate. Two years later, he presented the museum with bird skins on behalf of the Omani government, while also requesting a confirmation of the specimens' identification.[128] By accepting these specimens and offering identification services, institutions, such as the NHM, secured their position as centres of calculation in postcolonial times of budgetary austerity.

As European centres of calculation often competed with one another, they increased their demand for specimens from the Gulf and elsewhere. Markets for specimens and identification services thus emerged and became as important for Gulf biology as the oil markets. The Kew Royal Botanic Gardens and the Natural History Museum in London both claimed plants collected by British government expeditions, leading to 'cold wars' and open conflicts between the two institutions during the nineteenth century.[129] Only in 1951, during the period of postwar austerity, did both institutions agree on floristic 'spheres of interest'. In 1959, the British Treasury appointed a committee that led to an agreement according to which the NHM would deal with Northwest Africa, Europe, North America and the Polar Regions. Kew concentrated on the rest of Africa, Asia, South America and Australia. The two institutions subsequently exchanged parts of their collections.[130] However, as long as both institutions retained expertise in Arabia, Gulf biologists were able to deal with either one of them. Gulf scientists thus had a choice between different partners, reducing their dependence on any single centre.

While transnational networks between scientists and institutions inside and outside the Gulf were important, so were local networks. This book thus aims to tell a local as well as transregional story. In a continuation of medieval princely pastimes, many sheikhs practised game hunting, falconry or horsemanship.[131] In order to keep a reservoir of prey animals, they thus had an interest in breeding certain species and protecting them from extinction. Gulf rulers were also interested in environmental protection as a means to gain political legitimacy. Conservation scientists were able to benefit from these interests by gaining personal patronage from the Gulf's sheikhs. In this case, science in transaction involved an exchange of 'environmental credentials'[132] and expertise for funding and protection.

Beyond the elites, connections with wider society and local culture were also important. Without these connections, expatriates sometimes felt alienated in the Gulf and left well-paying jobs. The marine biologist John Burchard, for instance, was frustrated after working for Aramco during the 1970s. Although he had spent two winter seasons as a member of King Khalid's hunting party,[133] he did not feel comfortable in Saudi Arabia. While he accepted a new position at the University of Petroleum and Minerals in 1980, he confided to a friend at the NHM that he was 'getting rather fed up with fanaticism, dirt and political instability, quite apart from cultural, climatic and linguistic considerations'. By 1982, when his contract with the university would come up for renewal, Burchard hoped 'to have produced one or two solid works on reef ecology, and to have found a reasonable position in a more congenial environment'. In 1981, he again wrote to his friend that he did not 'plan to stay in Saudi indefinitely; just go, do a good piece of work or two and then get out'.[134] While some foreign scientists undertook innovative research even during short-term stays, high turnover rates of staff were costly for institutions in the Gulf.

However, a lack of connections with the wider Gulf society could also facilitate research on sensitive topics, such as evolution. Unfamiliarity with taboos sometimes led foreign researchers into areas avoided by others. Similarly, researchers who intended to stay in the Gulf countries for limited periods and to pursue a transnational career were less likely to practise self-censorship and potentially damage their reputation outside the Gulf. John Burchard, for instance, did not shy away from writing about evolution. An Aramco book on *Biotopes of the Western Arabian Gulf*, which he co-authored, stated that hamour predation on a shrimp population removed 'sick or unfit individuals from the population, thereby maintaining the overall standard of adaptation or fitness'. Burchard and his co-authors added that the outcome of 'true competition' between predator species over prey was 'increased adaptation or fitness of all the participants, or in other words, evolutionary progress'.[135] Gulf institutions interested in controversial areas of research thus benefitted from transnationally mobile scientists, even if they needed to replace them more often.

In investigating the transnational and local networks, this book uses a qualitative rather than quantitative approach. This means a focus on the lives and works of certain scientists rather than an effort to map the biological profession in the Gulf as a whole. The biographical focus also means that I will concentrate on a few strong connections rather than trying to illustrate all the networks of a Gulf scientist. This qualitative approach is appropriate, as the circulation and exchange of biological theories, such as evolution, are difficult to quantify. A focus on the quality of connections rather than their numbers also allows for a deeper investigation of the symbolic and broader cultural aspects of biological research. These aspects are especially important, as biologists in the Gulf monarchies often worked on animals, such as the Arabian oryx, that had high symbolic, but little agricultural value.

## Transnational collections and connections

As modern biology in the Gulf was transnational, so are the sources for its history. Based on their long-standing and strong connections with the Gulf, British institutions in particular contain a wealth of them. As governmental funding stagnated or decreased during the postcolonial era, institutions such as the Natural History Museum sought to diversify their sources of income. They thus considered not only their specimens and expertise, but also their archival collections, as valuable assets. They set up digitisation and document delivery services that sold copies of historical documents to researchers around the world. I am very grateful to the NHM for providing these services. On a much larger scale than my individual requests, Gulf institutions also financed the digitisation of British records. A collaboration between Qatar Foundation, the Qatar National Library and the British Library produced the Qatar Digital Library, for instance, which provided open access to the India Office Records and other British sources.[136] In the Gulf itself, stricter control of state records prevented the opening and digitisation of national archives to the same extent. However, university libraries still made their collections accessible to researchers from many different countries. A number of researchers also kindly shared parts of their personal archives with me.

More internationally dispersed than the archives were the publications of Gulf biologists. Scientists working in the Gulf published in Indian, American, British, French, Dutch, German and other journals. Many of these journals transcended single nations altogether, as they had multinational publishers, such as Elsevier, and international editorial and advisory boards. Biological publications from the Gulf also aimed at an international audience, as they were mostly in English and acquired by libraries in many different countries. Gulf institutions also paid for open access to certain periodicals, such as the *Saudi Journal of Biological Sciences* or *Tribulus*, the bulletin of the Emirates Natural History Group. In other cases, organisations such as King Saud University and the National Wildlife Research Center in Saudi Arabia established online repositories of publications by their staff. Research from the Gulf thus became accessible to readers around the world.

Besides contributing to debates and theories – including evolution – publications by biologists in the Gulf provide a lot of information about the authors' networks. Lists of publications reveal patterns of co-authorship. In their acknowledgements, biologists usually thanked collaborators, contributors, reviewers, funders and patrons. The methods sections provide further details about these collaborations. With the rise of laboratory science in the nineteenth century, field biologists often omitted 'circumstantial evidence'.[137] In order to save space and represent their work as being closer to experimental laboratory science, biologists also came to hide much of the social contexts of their work. However, accounts in popular magazines such as *Saudi Aramco World* compensated for the lack of social detail. Newspapers also interviewed prominent biologists and reported on major scientific discoveries and collaborations, often paying more attention to social and institutional relations than to the details of scientific work.

Because the academic journals articles were often aimed at international audiences, they also omitted local details. While acknowledging academic collaborators and funding agencies, scientists often left out names of amateur helpers, especially Bedouin rangers and guides. In addition, biologists in the Gulf rarely discussed political and religious challenges, especially in researching the sensitive topic of evolution. In order to overcome these limitations of academic publications, I also corresponded and conducted semi-structured interviews with biologists from various institutions. In addition, a few biologists painted vivid pictures of their obstacles and achievements in autobiographical accounts. The evolutionist Abdulaziz Abuzinada, for instance, published an entire book on his career at King Saud University and the National Commission for Wildlife Conservation and Development.[138]

Based on transnational and local sources, this book tells the history of modern biological research in the Gulf. It covers major disciplines and fields of the biological sciences and gives special attention to research on the important and sensitive topics of evolution and, to a lesser extent, climate change. The second chapter discusses botanical research, including the establishment of herbaria and surveys of Arabian vegetation. It argues that botanists contributed to the creation of plant kingdoms in two ways: literally, by writing floras of the Gulf region and, metaphorically, by enabling the Gulf monarchies to control the plants within their territories and to green the deserts. They also practised development biology in a double sense, using notions of the evolution of species to serve the development of countries. However, botanists were not merely tools for 'high modernist'[139] states, but also agents who warned against the destructive effects of rapid development on plant life.

The book's third and fourth chapters analyse the history of zoology and conservation in the Gulf. Chapter 3 discusses the connections sustaining the conservation of, and research on, Arabian oryx, gazelles and hamadryas baboons. They include connections between princes, European business and scientists. Similar to botanical research, zoological surveys rendered 'legible'[140] and controllable the Arabian animal kingdoms. While serving princely agendas, scientists, however, also investigated sensitive topics, such as evolution. The fourth chapter analyses ornithology and the conservation of birds. However, whereas Charles Darwin began his famous book, *On the Origin of Species*, with an account of the breeding of pigeons,[141] ornithologists and sheikhs in Arabia were more interested in houbara bustards, the main prey in falconry. Researchers on scientific islands of efficiency, such as the National Wildlife Research Center, studied the adaptations and survival of these animals and the threats posed by climate change.

The fifth chapter tells the story of palaeontological discoveries in the Gulf monarchies, especially those of primates. It argues that these discoveries relied on larger and smaller fossil economies, systems of exchange between oil companies, museums, tourism authorities and geological surveys. The chapter further illustrates the debates raised by fossils and their limits within the political and religious environments of the Gulf. The sixth and final chapter concludes by arguing that rentier states were not an obstacle to research and innovation. Despite the effects of a rentier mentality, scientists in the Gulf were able to conduct research that challenged not only academic, but also religious views. The characteristics of this rentier science include a clientelism that sustained islands of efficiency and freedom alongside more bureaucratic and less efficient structures. Chapter 6 also discusses the implications of this book's findings for the study of science in the contemporary Arab world. It argues that rather than lamenting the low output of Arab countries in terms of numbers of indexed publications and patents, we should appreciate the broader contributions of the Arab world, including its people and environments, in the transnational creation of knowledge. Finally, the chapter offers the prospect of future transnationalisation of biological research in the Gulf, comprising collaborations with researchers based not only in the Middle East, Europe and America, but also in Central and East Asia.

# CHAPTER 2

# PLANT KINGDOMS

In 1999, the Saudi Ministry of Agriculture and Water (MAW) published a bilingual Arabic and English book of 700 pages on the *Vegetation of the Kingdom of Saudi Arabia*. Its authors were the Pakistani scientist Shaukat Chaudhary and his Saudi colleague Abdul Aziz Al-Jowaid. In their introduction, Chaudhary and Al-Jowaid justified the publication of the book with an economic argument that contained an implicit criticism of Saudi Arabia's reliance on oil. They stated that the preparation of the book was 'initiated on the specific instructions of His Excellency the Minister of Agriculture and Water to create awareness of the vegetation, the precious but common, hardy yet delicate and the only renewable resource of the Kingdom'.[1]

*Vegetation of the Kingdom of Saudi Arabia* formed part of wider efforts by the Gulf governments and scientists to turn desert or oil monarchies into plant kingdoms. Oil and gas (the legacy of past life on Earth) still drove the economies of these kingdoms, but living plants also played important roles. The creation of plant kingdoms involved the 'greening' of countries such as Saudi Arabia and the United Arab Emirates through agricultural development, park landscaping and nature reserves.[2] The Gulf states thus used their abundant fossil resources to desalinate water, create agricultural subsidies and reshape environments for living plants, animals and humans. In 1973, Zayed bin Sultan, the president of the United Arab Emirates, invited a delegation by the Food and Agriculture Organisation (FAO) of the United Nations. Following their recommendations, he ordered the planting of over 4 million trees by 1981, with each one consuming three gallons of water, or $15, daily.[3]

Around the same time, the Saudi government started subsidising wheat agriculture on a massive scale, turning the kingdom into the world's sixth-largest wheat exporter by the early 1990s.[4]

The massive development of agriculture in Saudi Arabia during the 1980s was mainly aimed at diversifying the economy and creating self-sufficiency in essential commodities, such as wheat. However, Saudi subsidies for agriculture also had wider social and political objectives. They sought to distribute patronage and improve living standards for people in rural areas in order to prevent them from moving to the cities. Agricultural expansion also created national pride,[5] which the Saudi government used to generate a modern legitimacy based on science, in addition to its Islamic credentials.[6] In 1984, Saudi Arabia began exporting wheat to Europe, the Soviet Union and China. A book published by the Saudi Ministry of Education in 2000 listed this event among the 'hundred most important events in a hundred years', mentioning that the kingdom received a certificate of merit from the FAO for its 'achievements in the area of agricultural production'.[7]

While agricultural development had national aims connected to self-sufficiency and pride, it relied on global postcolonial expertise. As the Gulf states remained part of Britain's informal empire until the 1970s,[8] many of the agricultural scientists and botanists working in the region, such as Shaukat Chaudhary, came from Britain and its former colonies. The Food and Agriculture Organization itself continued the ideas of the British Colonial Advisory Council on Agriculture and Animal Health, and employed many experts who had previously served the British Empire.[9] While working for the Saudi MAW, Chaudhary served as an FAO plant taxonomist between 1989 and 2001.[10] However, in the process of decolonisation, the Gulf also became open to scientists from other countries and historical rivals of Britain in the Middle East, such as Germany. Decolonisation thus facilitated a further transnationalisation of Gulf science. This chapter will show that German, alongside British, Egyptian and American expertise were important for botanical and agricultural research in the Gulf.

Besides self-sufficiency and legitimacy, agricultural development and science served another goal of the Saudi state, that is to control nature and society.[11] By subsidising and controlling agricultural production, the state made large sections of Saudi Arabia's human, plant and animal populations dependent on the central government. When Abdullah ibn

Muammar, the minister of agriculture and water, announced cuts in wheat subsidies in 1994, he thus caused widespread discontent and unrest in the agricultural area of Buraydah.[12] Botanists and agricultural scientists played important roles in extending state control over plant life. One of their most important tools was taxonomy, the science of identifying, classifying and hierarchising organisms.[13] By producing taxonomical surveys and reference works, such as *Vegetation of the Kingdom of Saudi Arabia*, scientist made nature legible and subject to manipulation. Botanists thus created or contributed to plant kingdoms in a double sense. First, literally, they authored floras, descriptive catalogues of Arabian plants. Second, metaphorically, they provided some of the knowledge necessary for greening the desert monarchies.

However, while attempting to control their environments through science, the Gulf governments were unable or unwilling to control the scientists themselves entirely. Although they served the developmental aims of the Gulf states, many botanists and agricultural scientists retained considerable status and agency through their expertise and employment by foreign organisations. This allowed them to pursue research in more sensitive areas of botany, such as evolution. To some extent, their work on elements of the theory of evolution, such as adaptations, was even necessary. In order to expand agriculture and move organisms into new areas, the scientists needed to know about the organisms' adaptations to their environments and the conditions for their survival.

Agricultural and botanical research in the Gulf was also connected to evolution in a deeper way. The idea of development followed by the Gulf governments and agricultural scientists had itself strong historical links with a 'global Darwin.' Different readings of Darwin in Arabic as well as other languages connected evolution to notions of improvement, development and progress.[14] Common Arabic words for evolution, such as *taṭawwur, nushū'* and *irtiqā'*, also meant development and progress. Perhaps unaware of these strong historical links, many people in the Gulf still distinguished between economic development and Darwin's theory, embracing the former and challenging the latter. However, many botanists and agricultural scientists worked within the frames of both. In order to create plant kingdoms, they thus practised development biology, again in a double sense. They used notions of the development of species to serve the development of countries.

Combined with Islam, the products of development biology in the Gulf, such as *Vegetation of the Kingdom of Saudi Arabia*, sometimes contained a curious mix of creationism and references to adaptation. Abdullah ibn Muammar, the minister of agriculture and water, began his foreword to the book with a quotation from the Qur'an that referred to God's creation of the world: 'And We have spread out the spacious earth.' Ibn Muammar continued in his own words, 'Having created the universe with all its intricate cycles, forms, and ecosystems, God has made man, by virtue of his intelligence, a steward of this complex cosmic system.' Later in the foreword, the minister stated that 'man in the Arabian Peninsula had to adapt his way of life to its limitations and capacities, and he understood the importance of plant cover, as he obtained from it his food and extracted from it cures'.[15]

Ibn Muammar's brief mention of adaptation in the foreword did not explicitly contradict his creationism. Yet, the main text of *Vegetation of the Kingdom of Saudi Arabia* discussed the adaptation of plants in detail. In this discussion, the book's main authors, Shaukat Chaudhary and Abdul Aziz Al-Jowaid, used further important evolutionary concepts. They stated that 'for the organisms adapted for (properly equipped and prepared, and used to) survival in the desert environment, the desert is a haven providing them sanctuary from competition from other organisms as non-adapted organisms either not venture into the desert environment or they perish during such venture'. Adaptations to short and infrequent rainfalls included small size and the production of few flowers.[16]

The statements of Ibn Muammar, Chaudhary and Al-Jowaid do not necessarily reveal the tensions between creationism and evolution in Saudi Arabia. On the contrary, *Vegetation of the Kingdom of Saudi Arabia* could be seen as an example that creationism and certain elements of evolution, such as adaptation, could well co-exist in official publications. Indeed, a Saudi biology textbook for secondary schools from 2007 even defined adaptation in creationist terms. Adaptations, the textbook stated, were 'structural, functional and behavioral characteristics in organisms that help them to survive in their environment. Allah, glory to him, created for organisms those characteristics and structures that enable them to live in their different environments.'[17]

While also present in textbooks, the curious mixture of creationism and evolutionary concepts, such as adaptation, competition and survival in *Vegetation of the Kingdom of Saudi Arabia*, also reveals the transnational character of biological research in the Gulf. Written in both English and Arabic, the book aimed at local as well as foreign audiences. Co-authored by a Pakistani and a Saudi scientist, the book was a product of the transnational collaboration that characterised much of Gulf science. This chapter investigates these transnational networks with a focus on botany, one of the oldest and most important branches of biological research in the Gulf.

## Development biology

Even before the era of development during the second half of the twentieth century, Gulf governments were interested in research on plants as a means to increase agricultural production and achieve food security for growing populations. In addition, securing the land and food supply served the Saudi government in controlling the army and the population.[18] During the first half of the twentieth century, the Gulf monarchies especially feared locusts, whose swarms were migratory and could greatly damage crops. In order to benefit from international expertise on locust control, the Gulf governments reached out to foreign specialists. This occurred within the wider context of the internationalisation of locust control under the leadership of European colonial powers. Between 1931 and 1938, five international locust conferences were organised.[19] Saudi Arabia participated in the fourth of these conferences, which took place in Cairo in 1936.[20]

During World War II, disruption to shipping caused concerns in the British government about the food supply in the Middle East. Due to this situation, the British established the Middle East Supply Centre in Cairo in 1941. Fearing locust-induced famine in the Middle East, which would also threaten British control over the region, the centre launched a series of five annual anti-locust campaigns in Arabia. A small unit with British and Indian personnel operated in Oman and eastern Saudi Arabia between 1942 and 1943. Three succeeding operations included a large convoy that travelled from Cairo to Baghdad, Basra and Kuwait, and ended in Dhahran in eastern Saudi Arabia. After the closure of the Middle East Supply Centre at the end of World War II, the British remodelled the anti-locust campaigns on civilian lines and established the campaigns' new Arabian headquarters near Jeddah.[21]

During the first half of the twentieth century, the Gulf monarchies lacked the capabilities and expertise to organise such anti-locust expeditions. At the same time, the British Empire guarded the smaller Gulf states against operations by other foreign powers that could threaten its own influence in the region. The British-led expeditions were thus the only ones that were allowed to operate in the entire Arabian peninsula. Although botanical studies were not their main purpose, the reports of the expeditions included many floristic and ecological observations. Furthermore, the expeditions brought numerous dried specimens of Arabian plants back to Cairo. Botanists deposited about 6,000 specimens, representing over 750 species, in the herbarium of the Agricultural Museum in Cairo and placed duplicates in the herbarium of Cairo University.[22] Through a late colonial connection with Britain, Cairo thus received an important collection of plants from the Arabian peninsula.

Late colonial and postcolonial connections with Britain also enabled Egyptian botanists to build on their collections and develop expertise in plants from the Gulf. While British forces withdrew from Egypt at the end of the Suez Crisis in 1956, the specimens brought to Egypt by the locusts expeditions remained. Without any material for comparison, Egyptian scientists, however, found it difficult to identify the plants. They thus sought the help of Dorothy Hillcoat from the Natural History Museum in London, who had worked on Arabian plants since 1946 and was preparing *Flora Arabica*. Hillcoat stayed with Ahmed Khattab, one of the original members of the locust expeditions, and his colleague Nabil El-Hadidi, in Cairo for six weeks in 1969. She checked the determinations by her Egyptian colleagues and named other plants. By the end of her stay, only 50 plants out of the whole collection remained unnamed. With Hillcoat's help, Khattab and El-Hadidi were able to publish *Results of a Botanic Expedition to Arabia* as a book in 1971.[23]

From the 1950s onwards, the newly established universities in the Gulf were looking for experienced, preferably Arabic-speaking faculty members to build up their departments. Because of the Egyptian experience in Arabian flora, the universities were especially interested in hiring Egyptians to teach botany and undertake research on Arabian flora. The Egyptian government facilitated the migration of scientists to the Gulf, despite, and partly because of, the rivalry between Saudi Arabia and Egypt. During the 1950s and 1960s, the Egyptian government limited emigration in general in order to retain skilled workers for its development programmes. Yet, it allowed the migration of teachers and other professionals to the Gulf monarchies in order to display solidarity, promote pan-Arabism and increase its influence in the Arab world. Furthermore, the government of Gamal Abdel Nasser persecuted many educated members of the Muslim Brotherhood. In its rivalry with Nasser, King Faisal's government, which promoted Muslim solidary, granted many of these Muslim Brothers asylum and positions. The Egyptian–Saudi rivalry lessened during the early 1970s after the Egyptian defeats by Israel in 1967 and 1973, and Nasser's death in 1970. Yet, the economic difficulties created under Nasser led the Egyptian government to encourage large-scale emigration in order to gain remittances and reduce unemployment. In order to make temporary migration attractive for faculty members at Egyptian universities, the Egyptian government granted them the right to return to their positions after serving abroad.[24]

Even during a time of rivalry between Egypt and Saudi Arabia, a network of Egyptian botanists thus established itself at Gulf universities. One of the principal nodes in this network was Ahmad Migahid, whom Abdulaziz Abuzinada described as 'professor of a generation [ustādh al-jīl]'.[25] A graduate from Cairo University, Migahid led a research school in plant ecology and, in 1952, took over the chair of general botany at his alma mater. In 1965, he was seconded to King Saud University, where he taught until 1984. During the 1970s, he co-authored Flora of Saudi Arabia and, together with Abuzinada and others, edited the proceedings of two conferences on the Biological Aspects of Saudi Arabia.[26] In addition, Migahid built up a network of students working on Arabian vegetation. One of the most important of these students, and in many ways his successor, was another Egyptian named Kamal Batanouny. Migahid exerted a 'profound influence' on Batanouny,[27] who developed a 'real attraction to the desert' in an ecological field course taught by Migahid.[28] Serving as a faculty member at Cairo University, Batanouny had established his own school of arid environments with around 25 students by the mid-1980s.[29]

Batanouny further strengthened the botanical network between Cairo and the Gulf. Following Migahid, he was seconded to Gulf universities for years and wrote foundational works on Arabian environments. In 1975, Batanouny joined the Department of Biology at King Abdulaziz

University in Jeddah. As a basis for further research on Arabian environments, he published a bibliography entitled *Natural History of Saudi Arabia* in 1978. For this bibliography, the Egyptian botanist relied on his connections with Saudi and American colleagues. Batanouny received help from a young Saudi colleague in the department named Abdelelah Banaja. Abdulaziz Abuzinada, the president of the newly founded Saudi Biological Society, was also 'keen on having this bibliography' and encouraged Batanouny to complete it.[30] However, Egyptian and Saudi resources alone were not sufficient for Batanouny's ambitions. Around 1990, the libraries of the young Saudi universities were still rather small. In 1982, King Abdulaziz University held around 250,000 monographs. Even the largest Saudi university, King Saud University, held around 800,000 monographs at that time.[31] For this reason, Batanouny relied on the lists of publications that he received from the University of Arizona's Office of Arid Lands Studies.[32]

From Saudi Arabia, Batanouny moved to Qatar in the late 1970s, extending the network of his former teacher Migahid to the smaller Gulf states. In 1978, Batanouny became the chair of the newly established Department of Botany at Qatar University. There, Batanouny engaged in similar survey work, as Migahid did in Saudi Arabia. In 1981, Batanouny published a book entitled *Ecology and Flora of Qatar*. In this work, Batanouny relied on his local connections. The Qatari Ministry of Defence provided him with helicopters for four surveying trips. In addition, Hassan bin Mohamed Al Thani, a member of Qatar's ruling family, allowed him to collect plants from his private farm Al Wabra.[33] Batanouny also expressed his 'sincere appreciation' for the Bedouins, from whom he had learned 'more than anyone expects'.[34]

Besides local networks, connections with centres of calculation in Cairo and Kew facilitated Batanouny's work on the flora of Qatar. In Egypt, Nabil El-Hadidi, who had previously worked on the collections of the Arabian anti-locust expeditions, provided Batanouny with access to the Cairo University herbarium and checked his determination of plants. Batanouny also 'lavishly utilized' Migahid's reference work *Flora of Saudi Arabia*.[35] As Kew's herbarium possessed numerous specimens from Qatar,[36] Batanouny visited it for his book *Ecology and Flora of Qatar* and 'received every possible help' from its keeper and staff. In particular, Batanouny acknowledged the help of Tom Cope, who was responsible for grasses from South Asia and the Middle East.

Although Cope had not been to Qatar, his position at Kew allowed him determine many specimens of grasses for Batanouny.[37]

Egyptian botanists not only surveyed and described the vegetation of the Gulf monarchies; they also helped establish herbaria in the Gulf countries themselves. These local herbaria perhaps provided the Gulf monarchies with some independence from centres of calculation in Kew and Cairo, as their establishment often relied on connections with these centres. One of the main botanists involved in founding herbaria in the Gulf was Loutfy Boulos, another student of the 'professor of a generation' Ahmad Migahid. Boulos completed a degree in chemistry and botany at Cairo University in 1954, around the same time as Kamal Batanouny graduated. Subsequently, Boulos worked in the herbarium of the Egyptian Ministry of Agriculture and, in 1971, joined his alma mater as an associate professor of botany. During the 1970s, he also started visiting Kew regularly in order to make use of its library and herbarium. 'He could not stand the heat in Cairo', said Tom Cope, 'so he spent the summer in Kew working.'[38] Despite his aversion to heat, however, Boulos followed his Egyptian teachers and colleagues to the Gulf monarchies. In 1978, Boulos visited Qatar and collected more than 300 plant specimens, parts of which he sent to Kew and Cairo, while other parts he kept in Doha.[39] Six years later, in 1984, Boulos became a professor of plant taxonomy at Kuwait University, an institution that prioritised the preparation of specimens, surveys and the classification of local species.[40]

Supported by his connections with Egypt and Britain, Loutfy Boulos surveyed the flora of Kuwait, as his colleagues Migahid and Batanouny had done for Saudi Arabia and Qatar. Before coming to Kuwait, Boulos had worked for many years on the weeds of Egypt. In 1984, the year that he moved to Kuwait, Boulos and Nabil El-Hadidi published a book entitled *The Weed Flora of Egypt*. The positive reception of this volume encouraged Boulos to produce a similar survey for Kuwait. In 1987, Boulos invited Cope to join him in collecting plants in Kuwait. In 1988, the Egyptian botanist published a volume entitled *The Weed Flora of Kuwait*, relying mainly on a manuscript given to him by Tom Cope for the nomenclature of the grasses.[41] In 1990, on the eve of the Iraqi invasion of Kuwait, Boulos and Cope were still collecting plants in Wadi al-Batin near the Kuwaiti–Iraqi border. At one point, Kuwaiti soldiers almost arrested them. Fortunately, Egyptian expatriates staffed not only the universities of the Gulf, but also the military. One of the soldiers who sought to arrest the botanists was actually from Upper Egypt. Putting on his best Upper Egyptian accent, Boulos persuaded the soldiers not to arrest them but to drive him and Cope even deeper into the border region to collect plants.[42]

The collections by the botanists from Cairo and Kew subsequently formed major parts of the Gulf's herbaria. Kamal Batanouny contributed many specimens to the herbarium of Qatar University, which was established in 1986.[43] Besides describing the flora of Kuwait, Loutfy Boulos was responsible for major additions to the herbarium at Kuwait University. After his arrival at the university in 1984, he managed to receive collections of plants from all the countries on the Gulf Cooperation Council, plus Yemen. By the late 1980s, the herbarium consisted of around 22,000 specimens and was considered one of the most comprehensive collections of Arabian plants. However, in August 1990, Iraqi forces invaded Kuwait. Within days, the Kuwaiti forces were either destroyed or had withdrawn to Saudi Arabia. As the Iraqis occupied Kuwait, they moved the entire herbarium, together with the library and other scientific collections and equipment, to Iraq. Boulos was not able to rescue any parts of the herbarium, as he was at Kew for one of his usual summer stays when the invasion happened. After the war, Boulos and his colleagues had to rebuild the herbaria through new collections in the field. The Gulf War thus caused a setback for botany and science in general in Kuwait.[44]

With its own Arabian collections intact, the Kew Royal Botanic Gardens remained a centre of calculation. While Tom Cope collaborated with scientists in different Gulf countries, he was able to produce surveys of the Arabian peninsula as a whole. In 1985, Cope published *A Key to the Grasses of the Arabian Peninsula*.[45] In 1996, he co-authored, together with Tony Miller, a colleague from the Royal Botanic Garden Edinburgh, *Flora of the Arabian Peninsula and Socotra*. This work was supposed to rectify the 'absence of a modern Flora covering the entire region'. In their acknowledgements, the authors thanked Loutfy Boulos for his 'critical revision' of parts of the book and 'his support and help over the years'.[46] Through their cooperation with Gulf-based researchers such as Boulos, Kew and Edinburgh were thus able to remain central nodes in Arabian botanical networks during the postcolonial era.

The Egyptian and British botanists working in Arabia mainly restricted their work to surveying the local flora, making few references to evolution. In their *Flora of the Arabian Peninsula and Socotra*, Cope and Miller only mentioned plant evolution in response to desert environments a few times.[47] Cope did not remember ever having an extended conversation about evolution with his Arab colleagues. 'We discussed taxonomy', he said, 'but not the mechanism behind it.' Cope also thought that 'two out of three taxonomists in Arabia were probably creationists.'[48] At the time, few biology professors taught evolution at Egyptian universities, which was partly due to indirect pressure by religious scholars via television and sermons at mosques.[49] In his book, *The Weed Flora of Kuwait*, Boulos accepted that plants change, but did not elaborate on this fact in evolutionary terms. He stated that the weed flora is a 'dynamic flora', which is 'made up of native and naturalized species which continue to undergo changes'.[50]

In parallel to the connections with Egypt and England, the Gulf monarchies also used their relations with America in order to gain expertise on plants and agriculture. In 1975, an agreement between Secretary of State Henry Kissinger and Crown Prince Fahd bin Abdulaziz established the United States–Saudi Arabian Joint Commission on Economic Cooperation. After the Arab oil embargo against the United States in 1973, this agreement aimed at restoring the political and economic relationship between both countries. Under the agreement, US government agencies were to provide technical assistance to Saudi ministries in the fields of agriculture and water, science and technology, and industrialisation. This programme differed from other American development programmes in that the Saudis agreed to cover the entire costs. This freed US agencies from many of the budget and legislative constraints that were associated with other foreign aid programmes.[51] Over the course of 25 years, until 2000, the commission channelled billions of oil dollars back to the United States without congressional oversight.[52]

The US–Saudi Arabian Joint Commission on Economic Cooperation undertook important work in environmental development. This work included the establishment of the kingdom's first national park in Asir following American models in 1981.[53] In 1975, the commission also agreed on a project on agriculture, water resources and land management that had an estimated cost of $56 million over ten years. By 1978,

17 specialists from the US Department of Agriculture (USDA), 11 from the Department of the Interior and seven from the American University of Beirut were working on the project. Since the 1930s, this university had already been part of American scientific networks that aimed at agricultural development in the Levant.[54] The researchers of the commission were based at the Saudi Ministry of Agriculture and Water and its National Agriculture and Water Research Center. They were charged with a variety of tasks in order to develop the kingdom's agriculture, including assessments of agricultural potential, soil surveys, natural resource inventories, and range and forestry management studies.[55]

Similar to the connections between Cairo, Kew and the smaller Gulf states of Qatar and Kuwait, a network between Riyadh, Washington and Beirut thus emerged and established a herbarium in the kingdom. Abdul Mon'im Talhouk, an entomologist from the American University of Beirut, found that his team could not identify the plants that they were collecting. Talhouk thus contacted a colleague from the American University of Beirut, the Pakistani botanist Shaukat Chaudhary, who later co-authored the book *Vegetation of the Kingdom of Saudi Arabia*. Chaudhary accepted Talhouk's invitation and was seconded to the National Agriculture and Water Research Center. Soon after his arrival in Riyadh in 1978, Chaudhary founded the National Herbarium of Saudi Arabia as part of the National Agriculture and Water Research Center. He found the laboratory at the centre 'well-equipped' and reference books, such as Migahid's *Flora of Saudi Arabia*, available. However, as Chaudhary kept collecting and pressing plants, he realised that he did not have a place to store his specimens. He therefore asked his supervisor from the USDA to order 30 herbarium cabinets from America. This was the start of the kingdom's National Herbarium, for which Chaudhary worked for more than 20 years.[56]

In order to build up the National Herbarium, Shaukat Chaudhary, similarly to Boulos in Kuwait, relied on a network of collectors who provided him with specimens. These collectors included not only professional botanists, but also many amateurs, who thus made up the actors in Gulf biology. One of the most important amateur collectors was a self-taught British botanist named Sheila Collenette. She first came to the kingdom in 1972, accompanying her husband who worked for the Saudi Ministry of Petroleum and Mineral Resources. Over more than

20 years, Collenette collected more than 2,000 species of wild flowers in the kingdom. In 1999, she published descriptions of them in a large volume entitled *Wild Flowers of Saudi Arabia*.[57] In order to support further studies, she donated much of her material to Chaudhary's National Herbarium of Saudi Arabia.[58]

In theory, amateurs had more freedom to pursue their research interests than did the professional scientists employed by universities and governments. Yet, in practice, amateurs frequently relied on connections with governmental agencies and professional scientists for support. Initially, Sheila Collenette depended on the Ministry of Petroleum and Minerals, for which her husband worked. Ghazi Sultan, the deputy minister of mineral resources, authorised her to use a large network of exploration camps and field helicopters to reach inaccessible sites. After her husband's retirement from the ministry, Abdulaziz Abuzinada as secretary general of the National Commission for Wildlife Conservation and Development arranged visas and official cars for her visits to the kingdom and authorised his staff to accompany her.[59]

That she was a woman as well as an amateur was not an obstacle to Collenette's research. Even in 'a most masculine state', such as Saudi Arabia,[60] the field 'exhibited a borderland sociology and a frontier mentality' in blurring the boundaries between women and men.[61] In Saudi cities, Collenette encountered the strict gender segregation that had increased not as a mere consequence of tradition or conservatism, but with urbanisation and the expansion of state authority.[62] Visiting male Saudi colleagues at their homes, she often found their wives 'hidden away'. However, in the field, she wore trousers and long, loose shirts, as 'skirts would have been hopeless for climbing the mountains'. Once, when asked whether Collenette was a man, her driver responded 'half-and-half', which she found 'very apt'. While the Committee for the Promotion of Virtue and the Prevention of Vice sought to control gender mixing in the cities, another *hay'ah*, the National Commission for Wildlife Conservation and Development, even sent young Saudi men to accompany her in the field so that they would learn how to collect plants.[63]

Collaboration with amateur collectors like Sheila Collenette enabled professional botanists, such as Shaukat Chaudhary, to publish books on the kingdom's plants. The Saudi Ministry of Agriculture and Water, for which Chaudhary worked, had a special interest in the study of weeds as perceived obstacles to national development. In 1985, one official stated that MAW had 'brought about a revolution in agriculture in this country'. However, this 'marvellous achievement' also made weeds an urgent 'problem'. The legitimation of the Saudi government aside, Shaukat Chaudhary also saw the fight against weeds as a contribution to economic development in the wider region. He noted that 'in the Middle East, half of the effort of farming may be devoted to the battle against weeds'. Weed control would relieve farmers from the burden of hand weeding and allow them to send their children to schools 'in order to do more productive work in the future'.[64]

While the scientists working for US–Saudi Arabian Joint Commission on Economic Cooperation were mainly foreigners, they also started producing Gulf-specific knowledge. Instead of relying on Western manuals of weeds, the scientists produced specialised works on the weeds of Saudi Arabia and its neighbours. In 1983, Chaudhary co-authored A Manual of Weeds of Central and Eastern Saudi Arabia,[65] followed by a Weed Control Handbook for Saudi Arabia in 1985. One official justified the latter publication by claiming that 'climatic, soil and social conditions of Saudi Arabia and its unique weed flora combined with the most sophisticated agricultural technology being used here, precluded the use of weed control manuals available for other countries'. The official also hoped that the book would be of 'tremendous help to agriculture not only in Saudi Arabia but also to other sister countries in the Arabian Peninsula, especially the Gulf States'.[66] In 1987, Chaudhary co-authored a further work entitled Weeds of Saudi Arabia and the Arabian Peninsula.[67]

Despite the sensitivity of evolution, Chaudhary wrote about weeds and crops in the evolutionary terms of competition and survival. In his book Weeds of Saudi Arabia and the Arabian Peninsula, he stated that weeds possessed 'characteristics that give them the ability to survive and compete with an advantage over most agricultural crops'. The botanist further explained that in 'natural plant communities, the requirements of different plant species are rather different, and the interspecific competition not severe'. In a stand of crops, however, 'the requirement of each individual plant is the same as the next one and this intraspecific (intravarietal to be more precise) competition is intense. Weeds competing with the crop for water, nutrients, light and space aggravate this competition and can bring the crop yield rattling down'.[68]

When it came to controlling weeds, Chaudhary used evolutionary terms too. Thus, evolution did not remain a theory to explain the history of life, but also provided practical solutions for agriculture in the Gulf. In his *Weed Control Handbook*, Chaudhary stated, 'In nature, biological control has existed as part of the evolutionary process in the forms of predators and the prey, parasites and the hosts, herbivores and the plants and inter- and intra-specific competitions.'[69] In agriculture, one way of controlling weeds was to use 'smother crops'. These crops germinate earlier than weeds, have more developed roots, or are more efficient at photosynthesis than the weeds. 'With one or several of such adaptive characters', Chaudhary wrote, 'these crops will be better able to compete with the weeds for water, nutrients, light and carbon dioxide.'[70] That Chaudhary was able to refer to evolution in his texts was facilitated by a social network that consisted of people who saw evolution as a marker of being a 'scientist'. The two reviewers of the manuscript of *Weeds of Saudi Arabia and the Arabian Peninsula* were Abdulaziz Abuzinada and Sheila Collenette. Collenette found it 'hard to believe in anything other than evolution when you see what goes on, both with plants, how they adapt, and animals and birds'. She added that Chaudhary believed in evolution, 'because he was also a scientist'.[71]

Once Chaudhary had published his books about weeds, his references to evolution did not cause any problems for him. This was not despite the lack of public education about evolution, but because of it. As the educational context in the Gulf had not created a public attuned to evolution as a controversial topic, it precluded further controversies about it. 'People in Saudi Arabia at that time were not aware of what evolution was, really', said Chaudhary. As many Saudis lacked experience in debates about science and religion, he also found it easy to defend a secular view of science. 'Whenever I would get into discussion with them', Chaudhary stated, 'I would tell them that religion is a matter of belief, but science comes from the mind. It is a matter of fact.' The Pakistani scientist also told his conversation partners, 'You can believe in whatever you want to, but don't mix it up with science.'[72]

The Egyptian and American scientific networks in Arabia existed in parallel, and connections between them were sometimes difficult to make. Chaudhary did not restrict his works to weeds, but extended his interests to the entire flora of Saudi Arabia, which the Egyptians were surveying. Initially, he wanted to assist the Egyptian 'professor of a generation' Ahmad Migahid, who was then a professor at King Saud University. He found Migahid's *Flora of Saudi Arabia* very useful, but thought that it was 'not complete' and 'misidentified' some common plants. He therefore typed out a list of the plants with the wrong names and handed it to Migahid. When Chaudhary learnt that Migahid was preparing a second edition of the *Flora*, he handed him a further list of mistakes. However, when the second edition of the *Flora* appeared, Chaudhary was disappointed that Migahid had not implemented his corrections.[73]

While the American and Egyptian botanical networks did not connect very well, each of them retained a connection with Kew. Although few British botanists were active on the ground, Kew thus kept a central position in Arabian botany. After failing to collaborate with the Egyptian professor, Chaudhary continued Migahid's pioneering role in describing the plant species of Saudi Arabia. During the 1990s, after Migahid had left the kingdom, the minister of agriculture and water asked Chaudhary to write a new *Flora of the Kingdom of Saudi Arabia*. This book appeared in 1999 and was the product of Chaudhary's own work and the connections that he had built up over decades. Chaudhary again received help from the self-taught British botanist Sheila Collenette. He acknowledged that she was 'more than a friend of our Herbarium not only through donation of part of her collection but also the exchange of information and reverification of a lot of our material. Also, she has kindly gone through the manuscript and made a lot of helpful suggestions.' In addition, Chaudhary mentioned that he 'could always count on' the 'kindness and promptness in help' by Tom Cope of Kew and Tony Miller of Edinburgh.[74]

## Maps and mosses

While Kew and Edinburgh remained as botanical centres of calculation even after the formal British withdrawal from the Gulf in the 1970s, decolonisation opened up the Middle East for scientists from other countries too. They included scientists from Germany, a previous rival of the British Empire in Mesopotamia[75] and the wider Middle East. During the late nineteenth century, Germany followed Britain and the other imperial powers in establishing institutions for colonial botany.[76] In 1891, the German government founded the Botanische Zentralstelle für die deutschen Kolonien in Berlin, which was modeled after Kew,[77] and, in 1898, the Deutsche Kolonialschule in Witzenhausen near Kassel. In the colonies themselves, German scientists who were inspired by Darwin's discovery of natural selection also made important discoveries on plant evolution. In the tropics, they found plants surviving under extreme conditions of heat and moisture, which allowed them to recognise adaptations to the environment more easily than botanists working in Europe. Taking 'nature as the laboratory', German scientists thus established a Darwinian plant ecology in the late nineteenth century.[78]

Although Germany lost its colonies because of World War I, research on overseas flora continued. After the allies had initially expropriated German farmers, they allowed them to purchase land again during the 1920s and 1930s. The Deutsche Kolonialschule in Witzenhausen continued until 1944 and was re-opened as Deutsches Institut für tropische und subtropische Landwirtschaft in 1957. In 1971, the newly founded University of Kassel continued the teaching of tropical and subtropical agriculture at Witzenhausen. The German government dissolved the Botanische Zentralstelle für die deutschen Kolonien in 1920, but revived it between 1941 and 1943.[79] The Berlin-Dahlem Botanical Garden and Botanical Museum, where the Zentralstelle was located, also expanded its collections during the interwar years. By 1943, its herbarium was among the largest in the world, together with those of Kew, Leningrad and Paris. After World War II, the Berlin-Dahlem Botanical Garden and Botanical Museum became part of Freie Universität Berlin and continued research on plants in the former German colonies. During the 1970s, botanists from Berlin travelled to Togo, a former German protectorate, in order to survey the country's flora.[80]

In addition to surveying overseas flora and filling German herbaria, German botanists were interested in creating vegetation maps. Revealing the adaptedness of plants to certain geographical areas, these maps were important for colonial and postcolonial agricultural development. During World War II, German botanists produced a vegetation map of Africa as part of a manual of colonial science, but Allied bombing destroyed the entire edition.[81] In 1963, Deutsche Forschungsgemeinschaft (DFG, German Research Foundation) decided to fund Afrika-Kartenwerk, an interdisciplinary project to map parts of the African continent. It was supposed to provide 'a good foundation for all questions of development aid'.[82] Afrika-Kartenwerk focused on four regions, including areas formerly belonging to German East Africa, German Cameroon and the French-controlled Maghreb.[83] During the late 1960s, Deutsche Forschungsgemeinschaft started funding a similar project entitled Tübinger Atlas des Vorderen Orients (TAVO, Tübingen Atlas of the Middle East). Organised as an interdisciplinary collaborative research centre (*Sonderforschungsbereich*), TAVO lasted for more than 20 years, from 1969 until 1993. Complementing Afrika-Kartenwerk, TAVO aimed at producing maps of the region stretching from Egypt to Afghanistan. Specialists on the ancient Near East, who were dissatisfied with the scarcity of historical maps of the region,[84] initiated the project, but also involved natural scientists. A botanist from Tübingen named Wolfgang Frey co-edited a series of supplement volumes (*Beihefte*) to the maps, which also included research on the evolution of Middle Eastern flora.[85]

Under the umbrella of Tübinger Atlas des Vorderen Orients, German botanists came to work in the Gulf in parallel with the Egyptians and Americans. Between 1974 and 1980, Wolfgang Frey headed TAVO's botany section and involved his own students in research on Middle Eastern plants. During these seven years, he and his students concentrated their research on Afghanistan, Iran, Turkey and Jordan.[86] Scientists from Edinburgh and Kassel, both centres of colonial and postcolonial botany, helped them by determining plants.[87] As part of this research, one of Frey's graduate students named Harald Kürschner worked on mosses in northern Iran and the vegetation of Anatolia.[88] His experience with TAVO arguably helped Wolfgang Frey gain a professorship at Freie Universität Berlin in 1980, while Kürschner remained in Tübingen. Between 1981 and 1985, TAVO's botany section, now headed by Kürschner, directed its attention towards the Arabian peninsula.[89] Seeking a permit for fieldwork in Saudi Arabia, the German scientists wrote to several universities in the kingdom. Kürschner considered himself 'lucky' in that one of their letters addressed to King Saud University ended up 'on Abuzinada's desk'. In the name of the Saudi Biological Society, Abuzinada thus invited the German researchers to give a presentation on TAVO and their plans to map Saudi vegetation.[90] Arriving in Riyadh in 1981, Frey found himself to have 'stumbled into paradise'.[91]

Frey's presentation in Riyadh formed the starting point of a multi-year collaboration between botanists in Germany and Saudi Arabia.

Between 1981 and 1984, the TAVO botanists travelled to the kingdom five times in order to map different regions. In turn, one of their Saudi counterparts, Abdullah El-Sheikh from King Saud University, visited Berlin three times during the period to do research.[92] In 1982, during one of his visits, El-Sheikh also gave a lecture on 'Plant Life in Saudi Arabia' at Freie Universität Berlin.[93] Even non-Saudi biologists at King Saud University were enthusiastic about TAVO. This enthusiasm perhaps partly derived from the wish of Arab expatriates to secure the renewal of their temporary, but lucrative, contracts at Saudi universities through their involvement in long-term projects. A Sudanese botanist at King Saud University proposed to map Saudi Arabia at the very detailed scale of 1:50,000. This would have required far more work than the actual TAVO map of central Saudi Arabia, which had a less detailed scale of 1:1,000,000.[94]

With the help of Abdulaziz Abuzinada, the TAVO researchers gained considerable support from Saudi institutions and expatriate researchers and support staff in the kingdom. While Deutsche

**Figure 2.1**   Abdullah El-Sheikh (smoking shisha, centre) and Wolfgang Frey (right) somewhere between Mecca and Ta'if in 1981 (courtesy of Harald Kürschner)

Forschungsgemeinschaft covered the costs of travelling from Germany to Saudi Arabia, the Saudi Arabian National Center for Science and Technology (SANCST) funded expenses inside the kingdom. In 1981, at the height of the Saudi oil boom, the German scientists even enjoyed first-class tickets on Saudi domestic flights. Kürschner emphasised the particular role of Abuzinada in making this possible. Abuzinada was a 'manager type', who 'pushed' the project, said Kürschner, adding that 'nothing would have been achieved without him. The other Saudis only drank tea and performed official duties. The work was left to the expatriates.'[95] Besides Abuzinada, Frey and Kürschner relied on the Saudi botanist Abdullah El-Sheikh and his Egyptian colleague Ahmad Migahid, the author of the first *Flora of Saudi Arabia*. These two scientists led the initial excursions in 1981 and 'acquainted' the Germans with Saudi vegetation.[96]

Encouraged by their successes in Saudi Arabia, the TAVO researchers extended their mapping work to Oman. In the sultanate, connections with locally based experts were also important for winning support from

**Figure 2.2** Abdullah El-Sheikh (kneeling, centre) and Wolfgang Frey (right, with hat), pressing and sorting plants somewhere between Mecca and Ta'if in 1981 (courtesy of Harald Kürschner)

government agencies. As Oman developed its own higher education institutions decades later than Saudi Arabia, these experts were, however, mainly foreigners themselves. The Germans' main contact was the British government advisor and ornithologist Michael Gallagher, who had started to develop a National Herbarium of Oman in 1982.[97] In 1983, Gallagher supported a field trip by Wolfgang Frey and Harald Kürschner to Muscat. A second contact for the German scientists was another British naturalist and advisor named Ralph Daly, who voiced particular 'interest' in a map of a nature reserve in Oman.[98] Daly and Gallagher had previously organised a multidisciplinary Oman Flora and Fauna Survey in the 1970s.[99] Thanks to the assistance from Gallagher and Daly, a team led by Frey and Kürschner was able to map the vegetation of much of northern Oman. In return, the German scientists made available the doublets of their collection of Omani plants and contributed to Gallagher's National Herbarium.[100]

In addition to the maps, the TAVO researchers also published books and chapters on Saudi and Omani vegetation within the *Beihefte* series co-edited by Wolfgang Frey. For these publications, the scientists again relied on connections with the Royal Botanic Garden Edinburgh, whose staff helped identify 'some critical specimens'.[101] In 1985, Abdullah El-Sheikh and Frey published a book on the flora of central Saudi Arabia.[102] The following year, Frey and Kürschner authored a chapter on 'remnants of vegetation' in Muscat. Although Omani governmental support had made their research possible, they expressed criticism of the country's 'development'. The German scientists warned that the 'diverse vegetation' of the Muscat area was 'threatened with near destruction' by increasing urbanisation and industrial development. They thus proposed that the protection of parts of the area be in order to 'put a check on the increasing construction activity'.[103]

While studying Arabian flora in general, Frey and Kürschner had a special interest in bryophytes and their evolution. During the 1970s, Frey was already working on the evolutionary development of mosses in parallel to his work for TAVO. In an article published in 1977, he analysed the structure of the tissue system and the evolutionary history of mosses. Based on fossils, Frey concluded that the 'origin' and 'divergence' of the large groups of mosses occurred around 400 million years ago.[104] While mapping Arabian vegetation in the 1980s, Frey and Kürschner continued his research on bryophytes. Support from Gulf institutions thus did not restrict them in their research interests. After their first trip to Saudi Arabia in 1981, the two botanists started publishing a series of 'Studies in Arabian bryophytes'. In what they called 'the first records of bryophytes from Saudi Arabia', the Germans described six mosses and four liverworts, which they had collected in 1981. Unexpectedly, they found some of the bryophytes 'at extremely dry sites in the vicinity of Riyadh' in central Saudi Arabia. This led Frey and Kürschner to expect 'a considerable number of bryophytes' in the wetter western parts of the kingdom, which were influenced by monsoon rains.[105]

Their interest in Arabian bryophytes also led the German botanists to continue their research in the kingdom after completing their work for TAVO. As in the case of Shaukat Chaudhary or the Egyptian botanists, their connections to the Gulf monarchies thus lasted longer than a single project. For the continuation of their research, the Germans again gained support from Saudi governmental institutions. In their final report to Deutsche Forschungsgemeinschaft in 1985, Frey and Kürschner described their contacts with Abuzinada, Abdullah El-Sheikh and SANCST as 'very good'. Their Saudi collaborators apparently had a strong interest in extending the cartographical work over the entire kingdom in order to create a 'Vegetation Atlas of Saudi Arabia'. Funded by SANCST, El-Sheikh, Frey and Kürschner planned to survey the flora of Saudi Arabia on 20 maps at the more detailed scale of 1:500,000. Besides this new atlas, the Riyadh Development Authority invited Frey and Kürschner to study the Thumamah Nature Park north of Riyadh in 1984 and 1985.[106] In 1988, the National Commission for Wildlife Conservation and Development invited Frey to study the bryophyte flora of western and central Saudi Arabia.[107]

When collaborating with scientists in Arabia, Frey and Kürschner initially avoided the topic of evolution. Kürschner remembered that he had 'never talked about' evolution with his Saudi counterparts.[108] Considering evolution of 'no relevance' for the mapping of the contemporary vegetation, Frey too 'spoke about it very little'. However, Frey also 'avoided' the topic 'for pragmatic reasons'. This pragmatism had arisen from his previous experience in Iran,[109] where he had undertaken research during the 1970s. In Iran, the theory of evolution was also controversial, and textbooks avoided applying it to humans.[110]

Although Frey and Kürschner avoided the discussion of evolution inside Arabia, TAVO's publications in Germany referred to the theory. In 1985, Tübinger Atlas des Vorderen Orients sponsored the Symposium on the Fauna and Zoogeography of the Middle East in Mainz, Germany. The entomologist Abdul Mon'im Talhouk and 41 other scientists from Kuwait University and other institutions participated in this symposium. The editors of the proceedings, who included the ichthyologist Friedhelm Krupp, emphasised the Middle East's 'importance in zoogeography and evolutionary biology, owing to its position on the crossroads of the Palearctic, the Afrotropical and the Oriental realms'. At the same time, the editors lamented that the Middle East had 'never received the attention it deserves' in comparison with other zoogeographical transition areas.[111]

During the early 1990s, the Arabian research by Wolfgang Frey and Harald Kürschner experienced a hiatus. One of the reasons was financial difficulties in the Gulf monarchies resulting from lower oil prices and the costs of two Gulf wars between 1980 and 1991. Saudi Arabia first supported Iraq against Iran with $26 billion, and then spent $55 billion to drive Iraq out of Kuwait. During the same time, oil prices dropped from $40 per barrel in 1981 to $10 in 1986, and mostly stayed below $20 throughout the 1990s.[112] In this context, state spending on botanical research 'collapsed' between 1985 and 1987. This meant not only an end to first-class tickets on Saudi domestic flights for the German researchers, but also the departure of many expatriate researchers. The Egyptian 'professor of a generation' Ahmad Migahid, who had previously collaborated with the Germans, left King Saud University in 1984.[113]

Besides the financial difficulties in the Gulf monarchies, another reason for the hiatus in the 1990s had to do with Frey's reliance on German agencies, especially Deutsche Forschungsgemeinschaft, for his work inside Germany. While international networks supported botanical research, they thus also made it vulnerable to funding cuts in different countries. In order to pay for equipment and doctoral students, Wolfgang Frey had to follow the trends in German research funding. During the 1980s and 90s, the interests of German agencies turned away from the Middle East and towards the tropics. With the growth of environmentalism in Europe in the 1980s, the tropics became 'the topic', Frey said. In order to gain funding for new projects and graduate students, he thus turned to the rain forests of Africa and South America.[114]

While Frey moved from deserts to rain forests, his colleague and former student Harald Kürschner found another researcher based in Arabia to resume his collaboration. This researcher was Shahina Ghazanfar, a Pakistani botanist with strong connections to England. Holding a doctorate from the University of Cambridge, Ghazanfar became one of the first biologists hired by Sultan Qaboos University. Like many other Gulf universities, this institution comprised mostly foreign faculty members. Ten years after its foundation in 1986, the university had around 400 faculty members, 90 per cent of whom were expatriates.[115] A British zoologist named Michael Delany built up the university's biology department. Recruiting mainly among British universities, Delany sent a job advert to the University of East Anglia, where Ghazanfar's husband Martin Fisher was studying. SQU hired both Fisher and Ghazanfar, and Ghazanfar started working as an assistant professor there in 1987.[116]

What facilitated the collaboration between Harald Kürschner and Shahina Ghazanfar was that Sultan Qaboos University actively cultivated links with foreign universities. By 1997, Sultan Qaboos University had at least 15 memoranda of co-operation in research, teaching and the exchange of staff and students with universities in Arab countries, America, Canada, Britain, France and Japan. SQU, like other Gulf universities, had a special interest in securing places for its graduates on programmes at foreign universities. In addition, SQU linked itself to other regional universities through the Gulf Cooperation Council and the Arab Bureau of Education for the Gulf States and participated in annual meetings of university presidents, deans, librarians and registrars.[117]

In order to prepare students for graduate studies in Europe and America, Sultan Qaboos University largely followed Western academic models. Ghazanfar conceived Sultan Qaboos himself as 'very open' in his vision for the university. The language of instruction in biology was English, and Ghazanfar was 'discouraged' from speaking in Arabic with her students.[118] SQU had nominal arrangements for gender segregation, but men and women shared most of the university's facilities in practice.[119] Ghazanfar even took her female students on field trips. She considered this very beneficial, as some of the girls had never left Muscat previously.[120] As in the case of Sheila Collenette's excursions, Ghazanfar's field trips thus challenged gender norms in the Gulf monarchies.

While Sultan Qaboos University linked itself to Western academia, its professors also had to adapt to the knowledge of local students. During the first ten years of its inception, SQU admitted the top 15 to 18 per cent of students who took the final secondary school examination.[121] Still, the first two years of the five-year biology degree focused on consolidating the students' high-school knowledge.[122] Martin Fisher thus found his teaching to be 'at a fairly low level, more like the equivalent of English senior secondary school'.[123] Only after the first two years did the teaching approach British 'degree levels'.[124] Fisher also found that during the initial years, the English of most of the students was 'very poor'.[125] As Delany and Fisher did not speak Arabic, they relied on translators in the classroom.[126]

In contrast to British universities, the teaching of evolution was also restricted at Sultan Qaboos University. While following Western models, SQU was not secular, but had a religious as well as national mission. Its first objective was 'to prepare qualified generations of Omanis who are aware of the national Islamic heritage of the sultanate, and to deepen their belief in God and their loyalty to the nation and the Sultan'. In biology, this meant that the faculty members were 'advised, initially, not to teach anything on evolution'.[127] The foreign professors used *Biology: A Functional Approach*, a major British textbook at the time. In its original form, this book ended with a section on 'Ecology and Evolution'. However, the publications department of Sultan Qaboos University 'removed' this section before distributing copies among the students.[128] This had not been decided by the head of the biology department, who was himself British, but by a more senior administrator, perhaps an Egyptian.[129] Outside of the textbook, Ghazanfar still tried to teach her students evolution. She found that 'for plants it was never a problem', but relating to man and animals, 'the students never accepted it'. 'The Qur'an said that God created man', Ghazanfar added, 'you can't argue with that.'[130]

In their research, the biologists at Sultan Qaboos University had more freedom than in their teaching, but also less support, which encouraged reliance on foreign funders. There were 'no real restrictions', said Martin Fisher, 'we did what we wanted to do'.[131] Faculty members were encouraged to carry out research and to attend conferences as part of a faculty development programme.[132] Funding for research, however, was very limited. For that reason, Fisher found that 'not a lot of research' took place 'at the Department of Biology or in any of the other departments really'. According to him, 'the majority of people were not doing any research or were doing smaller, trivial things'. Connections with funders in Switzerland and Britain, however, enabled Fisher and his wife Shahina Ghazanfar to continue publishing.[133] The two scientists thus played a foundational role in Omani botany. Together with her husband, Ghazanfar spent ten years at Sultan Qaboos University, from 1987 to 1997. Besides teaching courses on botany, she built up a university herbarium in 1987 and helped curate the National Herbarium. Finally, she was involved in landscaping and designing the University Botanic Garden, which contained a plant nursery for teaching and research.[134]

Connections with Kew were crucial for Ghazanfar, as they were for botanists elsewhere in Arabia. After graduating from Cambridge and before coming to SQU, the Pakistani botanist taught in Nigeria during the 1980s. As an expatriate, she regularly received roundtrip tickets back to England,[135] which she used to visit Kew and access publications unavailable in Nigeria. Later, in Oman, she often corresponded with Kew's Tom Cope, whom she first met in Oman in 1993, as well as with other Western taxonomists.[136] They provided her with literature and helped her identify Omani plants. In 2009, Ghazanfar joined Kew as head of the temperate regional team, which was responsible for plant-related matters in Europe, the Middle East, East Asia, America, Australia and New Zealand.[137] Continuing the circulation of knowledge between Kew and Oman in her new position, Ghazanfar felt so indebted to Kew that she continued to help her former Omani students.[138]

Using their connections with Europe, Ghazanfar and Martin Fisher edited a book entitled *Vegetation of the Arabian Peninsula*. Despite censorship in teaching, the botanists had enough freedom to cover evolution in this book. For one of the book's chapters, Ghazanfar teamed up with the German botanist Harald Kürschner, who had led TAVO's botany section during the 1980s. Together, Kürschner and Ghazanfar wrote a chapter on bryophytes and lichens. They found around 200 species of mosses, liverworts and hornworts that 'appear to have evolved under similar ecological conditions' in the Arabian peninsula. The two botanists ended their chapter with a call for more research into the evolutionary biology of bryophytes and lichens. They stated that these plants are 'ideal' for the 'study of the ecological, morphological and physiological adaptations that enable organisms to survive and evolve in extreme environments'.[139]

Kürschner soon followed his own call for more research on bryophyte evolution. Collaborating with colleagues in the Gulf monarchies for years thus did not turn him away from evolution. In 1998, he and his long-term collaborator Wolfgang Frey published an article entitled 'Desert bryophytes: adaptations and life strategies against solar radiation and desiccation'. This article summarised and built on decades of research that Kürschner and Frey had undertaken on Arabian vegetation. Like the chapter that Kürschner had written with Ghazanfar, the article talked about the evolution of bryophytes. The Germans wrote that for bryophytes, deserts are 'by no means an "empty quarter"'. Instead, they found that 'some species of mosses and liverworts have evolved special ecological, morphological and physiological adaptations', which allowed them 'to survive and even further evolve under extreme desert conditions'. These adaptations include a special leaf surface and air chambers that enable the plants to tolerate heat and water stress. Frey and Kürschner argued that these adaptions were of 'crucial importance for life in the desert'.[140]

Kürschner's work in Arabia contributed to further thinking about long-term bryophyte evolution. A decade later, Kürschner summarised the 'macroevolution' of mosses and liverworts. In 2008, he published an article in the *Turkish Journal of Botany*, a periodical that explicitly aimed at publishing articles on plant evolution, despite controversies over evolution in Turkey.[141] In the introduction to his article, the German botanist wrote that 'bryophytes originated in the Lower Devonian' around 350 million years ago. One hundred million years later, the major groups had 'evolved and are represented by today's evolutionary lines'. However, during the Cretaceous, a 'co-evolution' with flowering plants and a great diversification of the 'modern lines' of mosses occurred.[142]

Even after decades of work, the research by the Germans remained far from exhaustive. At the end of his article, Kürschner indicated that Arabian bryophytes still required much more research. At the same time, he described these plants as being in particular danger from development. He thus also connected research on past evolution with future conservation and criticised the environmental engineering of Arabian governments. The botanist pointed out that many South

Arabian mountains remained unstudied and speculated that many taxa were still awaiting discovery. He warned, however, that the 'human impact on the forest remnants drastically increases'. South Arabia was being 'developed intensively' and even remote parts were 'opened up'. The transformation of the economy was leading to 'increasing forest destruction'. Kürschner thus feared that many 'remarkable floro-historical elements will vanish or become extinct in the near future'.[143]

## Oases of Oman

Even before Harald Kürschner published his warning in 2008, other researchers had started to work on plants in the Arabian mountains. During the late 1990s, another interdisciplinary research project involving botanists in Germany and Arabia started. As with Tübinger Atlas des Vorderen Orients, this project emerged at the intersection of the life sciences, area studies and archaeology. Moreover, like Kürschner and Frey's entry into Arabia, the impetus for this project came by chance and from a connection between a German researcher and a senior administrator from the Gulf. As with TAVO, this project also began in Tübingen, one of the main hubs of German research on the Middle East. The head of the project was Heinz Gaube, a Tübingen orientalist who had co-organised TAVO. In 1995, he led a group of students on an excursion to Oman.[144] Filled with 'sadness', the group saw a 'decline of old oases settlements' as a result of many Omanis moving from the countryside to the cities. At the end of their trip, Gaube gave a lecture at Sultan Qaboos University. The vice-chancellor who attended the lecture convinced Gaube that he should document the condition of the old oases.[145]

In the following years, Gaube set up an interdisciplinary project entitled 'Transformation processes in oasis settlements of Oman'. Like TAVO previously, its main funding between 1999 and 2007 came from Deutsche Forschungsgemeinschaft. The project included the University of Tübingen as well as Sultan Qaboos University, the University of Stuttgart and the German Archaeological Institute.[146] However, rather than including Harald Kürschner and Wolfgang Frey from Freie Universität Berlin, Gaube recruited another German agricultural scientist named Andreas Buerkert. The project on Omani oases thus gained its own agricultural and botanical sections – as the TAVO centre had done before.[147] Buerkert worked at the University of Kassel at Witzenhausen, which had previously hosted the Deutsche Kolonialschule. The oases project thus formed a continuation of German colonial and postcolonial botany and agricultural science.

Like the foreign biologists working at Sultan Qaboos University, Buerkert had to adapt to the academic culture of the Gulf. Initially, he complained about cultural differences and a lack of critical thinking among SQU students.[148] Buerkert perceived SQU to be mainly 'an institution for the education of young girls and boys to remain on the path of virtue'. To him, it was 'not a university in the Western sense, in which critical questioning was encouraged to the advancement of science'.[149] The expatriate faculty members at SQU held similar views. They regarded the memorisation strategies prevalent among Omani students as being 'deficient' and sought to introduce problem-solving strategies instead.[150] One of these faculty members was Annette Patzelt, another German botanist and one of Buerkert's partners at Sultan Qaboos University. She had succeeded Shahina Ghazanfar at SQU in 1998 and taught biology there until 2006. Patzelt thought that 'the very good students were absolutely equal to the very good students in Europe'. However, overall, her Omani students not only 'had problems with the English language', but their 'analytical thinking, their scientific thinking was in need of strong encouragement'.[151]

One of the differences between the German and Omani academic cultures was that during the oases project in the 2000s, Sultan Qaboos University still lacked systematic teaching of evolution. Annette Patzelt stated that evolution was not even part of the general biology degree course. Only briefly did she mention to her students that, through evolution, new species of endemic plants emerged. The German botanist was unaware, whether the theory as such was 'taboo', but evolution was not part of the curriculum. Yet, Patzelt claimed students interested in evolution could have 'read about it' independently.[152]

During his first fieldwork trips, Buerkert also encountered the mistrust of oasis inhabitants and political sensitivities, which added to the difficulties in connecting German to Omani botany. While Buerkert's team was mapping remote oases by means of a Global Positioning System (GPS), some local inhabitants initially held them as spies. The German botanist also found irrigation to be a sensitive topic. He stated that 'modern drip irrigation' was 'the political aim' in the

Batinah region of northern Oman and that the traditional *falaj* irrigation system that relied on sloping tunnels was 'out'. Initially, the German researchers found it difficult to report that the traditional system had 'ecological advantages' and was unexpectedly efficient. They also found it sensitive to write that the officially favoured drip irrigation was 'technically difficult, very expensive, and associated with long-term soil salinity problems', issues that later became well recognised within Oman.[153] Rather than merely being a tool of an authoritarian 'high modernist' state,[154] the German scientists were aware of problems in governmental development strategies.

Despite the cultural and political sensitivities, the German researchers managed to connect with the Omani authorities. Although he faced 'bureaucratic' obstacles, Buerkert eventually gained 'understanding and support' from Omani institutions. After waiting for four years, Buerkert managed to attain a material transfer agreement, which allowed him to take the genetic material of endemic plants outside the country. This enabled him to study the material jointly with other German, Omani, Belgian and Russian scientists and to preserve the germplasm of Omani wheat and bananas in international genebanks. Important for the support from the Omani authorities was the involvement of Omani graduate students, who were able to organise helicopter flights by the Royal Air Force of Oman as well as other forms of help.[155]

The German researchers not only gained logistical support from the Omanis, but also included them in the production of research papers. From the start, Buerkert's team sought to recruit Omani doctoral students. As Oman had many 'nouveaux riches' at the time, Buerkert found it 'difficult' to attract Omani nationals. Many of those academically qualified were not prepared 'under the conditions of an international PhD programme' and to accept the associated salary level.[156] However, despite a widespread 'rentier mentality',[157] an Omani student named Sulaiman Alkhanjari displayed much 'academic interest' and put himself forward as a student. In 2002, he began a doctoral research project on the diversity of Omani wheat races. Under the supervision of Karl Hammer, another German botanist, and Buerkert, Alkhanjari enrolled at the University of Kassel.[158] Very interested in plant evolution, Karl Hammer edited the journals *Genetic Resources and Crop Evolution* and *Plant Systematics and Evolution*.[159]

Despite cultural differences, Alkhanjari became a very productive member of the German-led research team. Buerkert considered Alkhanjari an ideal collaborator, who was willing to spend many days working in the field. He contrasted Alkhanjari with many 'nouveaux riches' in Oman who 'all too often left physical work to expatriates'. The German scientist explained this exception from the rentier mentality with Alkhanjari's cosmopolitan background. From a Zanzibari family, Alkhanjari completed much of his education in Egypt and Iraq at a time when Oman did not have a comparable educational infrastructure.[160] In 1987, Alkhanjari gained a bachelor's degree from Al-Azhar University in Cairo and went on to complete a master's degree at the University of Nebraska.[161] Alkhanjari's German supervisors perceived his cosmopolitan background as a reason why the theory of evolution was not a point of debate with him. His main supervisor at the University of Kassel, Karl Hammer, imagined possible opposition towards work on 'the origin of man', but for Alkhanjari's research on plants, evolution was no 'problem'. Probably because Alkhanjari had studied in different countries, Hammer found his Omani student 'well acquainted' with, and not 'hostile' to, the theory.[162]

Supervised by Hammer and Buerkert, Alkhanjari used evolution to explain the diversity of Omani wheat races. He published several articles from his doctoral work in Hammer's journal *Genetic Resources and Crop Evolution*. In one of these articles, Alkhanjari wrote that Omani wheat races, after their introduction from Mesopotamia, 'experienced evolutionary modifications resulting from natural selection and adaptation to the harsh desert environment'.[163] In another article, the Omani botanist began with a more general point, stating that 'the evolutionary processes leading to the development of wheat landraces depend on various factors'. These include 'natural and artificial selection, domestication history and several thousand years of adaptation to cultivation environments'. This history, Alkhanjari claimed, made Omani wheat races 'one of the invaluable heritages that traditional farmers have given to the world'.[164]

As was the case with Harald Kürschner, Alkhanjari connected his research on past evolution with a concern for future preservation. He also showed the negative effects of the country's development on its plant life. In his doctoral dissertation, completed in 2005, he wrote that Oman's 'modernization' had resulted in a 'rapidly progressing genetic erosion'. He thus called for the preservation of the 'culturally unique, ecological niche environments which led to the evolution of the ancient wheat landraces'. He proposed policy efforts for conservation of these races within old oases themselves. 'For security reasons', he also suggested transferring duplicates of the genetic material collected during the project to an international genebank 'to ensure its future use as a world-wide heritage'.[165]

The German–Omani botanical connections remained strong even after the formal end of the oases project in 2007. After the completion of his doctorate in 2005, Alkhanjari continued to collaborate with his former supervisors Karl Hammer and Andreas Buerkert. In 2009, they published another article combining their interests in evolution and conservation. Based on an examination of the variation in several crop species, the botanists found that Oman had been 'an important country for the evolution and diversification of economic plants for a long time'. The scientists even compared the 'evolutionary power' of Oman to that of south Italy, arguing that it resulted from the trading connections of coastal Oman and the relative isolation of its mountain oases. Oman's position at the crossroads of sea routes attracted plants from southern Asia and from the eastern Mediterranean. At the same time, the isolation of Omani oases by steep mountains helped to conserve these different plants and promoted their separate 'evolution'.[166]

Although the Germans were the main funders and leaders of the oases project, the dissemination of its results was not restricted to European audiences, but also reached Oman. In order to convince the Omanis to preserve oases and wheat races, the scientists organised an exhibition entitled *Oases of Oman*, which moved through different parts of the sultanate in 2007, and focused on the threats to traditional settlements. Promoting the exhibition, Andreas Buerkert gave an interview to the *Oman Tribune*, an English-language newspaper in the sultanate. Buerkert again warned about the negative effects of rapid development, arguing that the connection of the Omani mountains 'with the modern transport system, competition with the world market and lack of well-trained labour would lead to the extinction of old plants in the oases'. The German scientist proposed that 'Omani oases should be protected and farmers' livelihoods strategies strengthened.' In order to save the traditional settlements, he suggested the introduction of 'certified eco-agriculture' and 'eco-friendly tourism in small groups'.[167]

The interest of the German–Omani team in conservation also led them to raise awareness about the effects of climate change. In a lecture at Sultan Qaboos University accompanying the *Oases of Oman* exhibition in 2007, Buerkert speculated about the effects of climate change on fruit production in high-altitude oases. He explained that pomegranates, apricots and walnuts needed at least 100 'hours of chilling' per year. However, with climate change the number of chilling hours had dropped in the Omani mountains. As a result, some farmers had seen their pomegranate production dramatically reduced.[168]

In their research and outreach work, Buerkert and his colleagues were also able to gain the support of the Omani Ministry of Environment and Climate Affairs. In 2010, they published a book entitled *Oases of Oman*, which Hamoud al-Busaidi, the minster, launched in Muscat in the presence of other officials and diplomats. In his foreword to the book, the Minister rehabilitated the *falaj* system, stating that it had 'made it possible to grow crops and fruit trees, such as date palm and citrus, nurturing the development of early settlements and civilisation throughout Oman's history'. The Minister also embraced plans to preserve the old oases, stating that the book 'aids in developing sustainable strategies for the future of our traditional settlements'.[169]

Published and launched in Muscat, the book *Oases of Oman* underwent censorship, but for some reason censorship did not target evolution. This allowed the botanists to make similar arguments in their book published inside Oman, as they also did in their articles in international journals. A chapter co-authored by Sulaiman Alkhanjari, Andreas Buerkert and Karl Hammer described the introduction of tropical crops, such as black plums, to Oman from southeast Asia and India. In Oman, these crops underwent 'further evolution', as they interacted with Omani races and evolved 'specific adaptations'. In the conclusion to the chapter, the German– Omani team repeated its earlier arguments about the 'evolutionary power' of Oman. The scientists argued that this power resulted from the availability of plants from different parts of the world and 'the conservation and evolution of plants in oases' isolated by steep mountains.[170]

This interdisciplinary connection with archaeologists also led the oases researchers to move beyond plant evolution into the realm of human cultural evolution. *Oases of Oman* comprised another chapter co- authored by Buerkert and a member of the German Archaeological Institute. Combining evolutionary biology and climate change research, this chapter was entitled 'From Hunter-Gatherer Communities to Oasis Cultures: Climate Change and Human Adaptation on the Oman Peninsula'. The chapter argued that oases represented 'a particular form of human adaptation to a desert environment' and thus formed part of the 'cultural evolution and survival strategies of human communities' in Arabia. Buerkert and his colleagues suggested that 'radical climatic and environmental changes' were responsible for changes in the local economic basis and 'adaptive social behaviour'.[171] These climatic changes included increases in rainfall over the millennia, which allowed for a transition from semi-nomadic pastoralism to rain-fed agriculture.

## Conclusions

Fuelled by oil and the ideology of progress, which itself had strong links to concepts of evolution, the Gulf monarchies funded initiatives in development and environmental engineering. These initiatives included research on plants intended to increase agricultural production and gain food security. From the 1950s onwards, scientists based in the university departments of botany and research centres in the Middle East and Europe systematically collected plants and produced comprehensive taxonomic works, manuals for farmers and vegetation maps covering different Gulf countries. While aiming at economic development, these botanical works also increased state capacity and power. Through classification and mapping, the Gulf states attempted to render legible and control the local plant kingdoms and the people who depended on them.

Transnational and local networks formed the basis of this expansion of state power over the Gulf's plant life. Most of the major botanists engaged in surveying vegetation were foreigners, including Egyptians, Pakistanis and Germans. They came to Arabia and gained the support of local institutions through intergovernmental agreements and well-connected individuals. In Arabia, the botanists also benefitted from an exchange with colonial and postcolonial centres of calculation, such as the Kew Royal Botanic Gardens. These centres benefitted from specimens sent by colleagues based in the Gulf, and in turn provided identification services and access to literature. Finally, networks extending beyond the discipline of biology were important, such as collaborations with archaeologists, Orientalists and amateur collectors and researchers.

Although partly serving expanding developmentalist states, the networks still allowed for considerable agency among scientists. Despite the sensitivity of evolution, scientists repeatedly used the theory and associated concepts, such as adaptation and competition, in their studies of wheat, weeds and mosses. They also warned about the threats to plant life posed by modernisation and construction activities. They thus criticised the developmentalism that had encouraged Gulf monarchies to fund biological research in the first place. Analysing the effects of climate change on plants, scientists implicitly also pointed to the dangers of the 'natural unsustainability'[172] of the Gulf states.

However, what also facilitated evolutionary studies on plants was the fact that evolution was less controversial in relation to plants than it was to animals. At Sultan Qaboos University in Oman, evolution was not part of the curriculum, but botanists were still able to teach their students about plant evolution. The next chapter looks at research on mammals, including primates, in the Gulf monarchies. It investigates the extent to which evolutionary studies were possible on species much closer to humans.

# CHAPTER 3

# SULTANS, CONSULTANTS AND CONSERVATIONISTS

In 1979, an unusual rescue operation took place at one of Rome's airports. Around 50 animals, including zebras and antelopes, were stranded on a plane that had arrived from South Africa. The animals were destined for a safari park, but the Italian authorities refused clearance for the plane to continue its journey from Rome to Naples because of 'inadequate documentation'. In the meantime, the heat in the plane's hold and the stench of ammonia from the urine caused respiratory difficulties in the animals and the crew. The crew and their assistants had blood-shot eyes and were barely able to talk due to breathing in the ammonia.[1] As the animals started dying within their confined quarters, the airport authorities debated what to do about them. The plight of the animals soon reached Hamad Al Khalifa, the Crown Prince of Bahrain, who offered to bring the animals to his country. However, by the time the Italians allowed the Bahraini government to act, 40 animals had died. After the Italians had cut through the red tape, the surviving animals were transferred to another plane, which flew them directly to Bahrain.[2]

In Bahrain, the animals were taken to Al-Areen Wildlife Park, which had been established a couple of years earlier, in 1976. It was a project of Crown Prince Hamad and one of the oldest modern conservation institutions in the Gulf. When the animals from Rome arrived at Al-Areen, a wave of letters from around the world thanked the Emir and the Crown Prince of Bahrain for the rescue. 'We were touched by such a response', Al-Areen's manager commented. 'It was especially satisfying because what we had been saying to international wildlife organizations – that Bahrain and the Arab world were serious about conservation – was proved correct.' The manager added, 'It was a tragic story that had a bit of satisfaction at the end.'[3]

The story of the Rome rescue exemplified that princes in the Gulf countries often had a great personal interest in wildlife conservation. This interest was perhaps greater than their interest in plants, with which the ministries of agriculture and university departments of botany were concerned, but less so the princes personally. The greater personal interest perhaps reflected a special cultural appreciation of animals in Arab culture. In hierarchies of beings, such as the one produced by the tenth-century Brethren of Purity (Ikhwan al-Safa), animals typically ranked above plants. In Arabic poetry, animals, and particularly large mammals like camels and gazelles, received much more praise than plants did. In medieval as well as modern times, many Arabs gave their children names of animals, such as *usāmah*, 'lion', and *fahd*, 'cheetah'.[4] For centuries, 'charismatic megafauna' also formed part of an 'animal economy' of gift exchange.[5] The Abbasid Caliph Harun al-Rashid, for instance, gave an elephant and monkeys to Charlemagne.[6] In 1917, Abdulaziz Al Saud, the emir of Najd, presented British agents with two Arabian oryx, a male and a female, as a gift for King George V. After lodging in the Zoological Gardens at Bombay, where the male died, the female arrived in London in 1920.[7]

As part of their general interest in charismatic megafauna, many Middle Eastern rulers established zoological gardens. Around 2,500 BCE, Egypt became home to the world's first zoo.[8] Muslim caliphs also kept wild animals in 'enclosures' (singular *hayr*) that partly formed hunting reserves. The Umayyad Caliph Abd al-Malik ibn Marwan kept lions at his court, and his son Hisham built two castles with enclosures, Qasr al-Hayr al-Gharbi and Qasr al-Hayr al-Sharqi.[9] In 1957, King Saud bin Abdulaziz established the Riyadh Zoo, the first modern zoological garden in the Arabian peninsula. In the 1980s, the Riyadh Zoo was demolished and rebuilt at a cost of more than $30 million. 'The main aim of the zoo is to conserve and breed the native species of Saudi Arabia', an official stated around 1990, 'and we're succeeding in doing that.' By that time, the zoo possessed striped hyenas, hamadryas baboons and other native species. It had also acquired a male Arabian oryx from Al-Areen Wildlife Park and was looking for a female one to start a breeding programme.[10]

In addition to *ḥayr*, officials in the Gulf also saw modern wildlife parks as continuing another traditional institution called *ḥimá*, 'protection' or 'sanctuary'. *Ḥimá* was a conservation system in which a piece of land was protected from overgrazing in order to allow for the flora and fauna to be replenished. These protected areas had existed in the Arabian peninsula since pre-Islamic times, and Saudi Arabia as the largest Gulf monarchy had around 3,000 *ḥimá*s in the middle of the twentieth century.[11] Yet, at around that time, the *ḥimá*s started to disappear. In an effort to strengthen central authority and weaken the tribes, King Abdulaziz abolished tribal boundaries in 1932. A further decree in 1953 abolished certain *ḥimá*s that Abdulaziz had established for the cavalry and government purposes. Yet, local emirs interpreted this decree as the opening of all rangeland for general use, resulting in the rapid overgrazing of many *ḥimá*s and soil erosion.[12]

Colonial and postcolonial interests helped revitalise traditional interests in conserving the environment in the Arab world. In North Africa, the French colonial administration used environmentalism as a means of social control over pastoral nomads. Through national parks and environmental laws, it restricted the nomads' mobility and access to resources.[13] The postcolonial Moroccan state, which was also interested in controlling the nomads, expanded many colonial national parks and created new ones.[14] After World War II, the British government spread environmentalism throughout the Middle East. In an attempt to salvage declining power, the British Middle East Office sought to create an informal empire in the region through environmental development initiatives. Its staff thus toured the Middle East and laid the blueprints for environmental development in countries such as Saudi Arabia and Bahrain. In this way European experts ran environmental and other affairs for Arab rulers even decades after formal decolonisation.[15] A disguised form of 'green imperialism' thus continued even in the Arabian peninsula, where colonial rule had been less direct than in many parts of Africa and South Asia.[16]

The presence of networks of Western experts contributed to an environmental rhetoric among officials in the Gulf that mirrored global discourses. In 1993, Abdulbar Al-Gain, the president of the Saudi Meteorology and Environmental Protection Administration (MEPA), stated that man finds himself at a point in history that is pivotal in terms of the very nature of his relationship to the natural world. His footprint is to be found everywhere throughout the Planet, in the air, in the deep seas, the forests and the polar ice. Human activities over the last century have so affected natural processes that the very atmosphere upon which life depends has been altered.

Al-Gain thus stressed the need for 'environmental sustainability'.[17] This rhetoric contrasted sharply with the 'natural unsustainability' of the Gulf monarchies, which consumed huge quantities of energy and water per capita.[18]

However, while the officials in Arabia adopted elements of the global environmental discourse, they also sought to localise them. As the Gulf states attempted to 'Islamise' modern science, they also sought to create an 'Islamic environmentalism'. This kind of environmentalism formed the ideological basis for the MEPA.[19] In 1983, MEPA commissioned a group of scholars at King Abdulaziz University to formulate an Islamic policy on the environment.[20] This policy appeared in two editions published by MEPA in association with the International Union for Conservation of Nature (IUCN).[21] Several of the scholars behind this paper subsequently acted as advisors to MEPA. Although they cited *ḥimás* as a precedent for Islamic environmentalism, they complained about the difficulties of implementing or preserving these traditional institutions in Saudi Arabia.[22]

The Islamic environmentalism adopted by MEPA encompassed creationism. The second edition of the MEPA–IUCN paper, entitled 'Environmental Protection in Islam', began by stating that 'all things that God has created in this universe are created in due proportion and measure both quantitatively and qualitatively'. The paper by MEPA and IUCN combined this creationism with the notion of man's stewardship over nature, stating that 'man is also the executor of God's injunctions and commands'. The paper continued that man ought to 'manage the earth in accordance with the purposes intended by its Creator; to utilize it for his own benefit and the benefit of other created beings, and for the fulfilment of his interests and of theirs'. Providing an Islamic justification for conservation, the scholars also stated that 'the right to utilize and harness natural resources, which God has granted man, necessarily involves an obligation on man's part to conserve them both quantitatively and qualitatively'.[23]

Despite its potential for Islamic and creationist interpretations, conservation could be connected with evolution too. The conservation of nature could allow some forms of life to 'to remain in existence in their natural state, evolve as have their ancestors before them throughout evolutionary time'. Protecting the wild relatives of domesticated plants and animals from extinction could also preserve 'evolutionary reservoirs, the raw materials for future adaptation'.[24] Alternatively, some people perceived zoos as attempting to avoid the evolutionary struggle for life. In 1961, *Aramco World*, the magazine of the Arabian American Oil Company, published an article about the Riyadh Zoo, stating that 'in the zoo there is no rule of survival of the fittest – no fear, no quick, violent death, no overwhelming pangs of hunger, no competition when seeking a mate'.[25]

As the Gulf monarchies funded institutions for biological conservation, such as zoos, during the second half of the twentieth century, they also imported and employed many zoologists and conservationists. While many of these professionals probably had evolution as part of their education, this chapter pays special attention to instances whereby evolution became an explicit part of their work. The chapter also investigates to what extent the patronage by the princes for conservation imposed constraints on, or provided protection and freedom for, work on potentially sensitive topics, such as the evolution of ungulates and primates.

Going beyond university departments to trace transnational networks involving zoos and museums, this chapter takes inspiration from Lynn Nyhart, who has studied the history of biology with a focus on civic institutions in nineteenth-century Germany. Because of conservative fears about evolution, the Prussian education administration decided in 1882 to remove the subject of biology from the upper years of the university preparatory schools, the *Gymnasien*. In the context of this controversy, German naturalists developed a 'biological perspective' that focused on adaptation rather than the ultimate causes of evolution.[26] Like nineteenth-century German naturalists, and arguably inspired by European ecology more broadly, biologists in the Gulf also tended to focus on adaptations and be silent or creationist on the question of ultimate origins. However, they were acutely aware of the possible endpoints of evolutionary history: extinctions.

## Operation Oryx

After World War II, the oryx became a symbol of international conservationists and the Gulf monarchies alike. Because ancient Egyptians had bound its two horns together, or because the two horns looked like one from the side, the oryx was associated with the mystical unicorn.[27] As it was one of the rarest animals, the Fauna Preservation Society in 1950 named its journal after it. This society had roots in European colonialism and was founded by British conservationists in Africa in 1903 as the Society for the Preservation of the Wild Fauna of the Empire.[28] In the Gulf, the Arabian oryx was a symbol of beauty. Many parents gave its Arabic name, *mahā* or *al-mahā*, to girls. During the second half of the twentieth century, the Arabian oryx even became a national symbol and 'cultural icon'[29] of the Gulf monarchies. Its head appeared on the planes of Qatar Airways, the newspaper *Times of Oman*, and the 50-dirham notes of the United Arab Emirates.[30] Yet despite, or perhaps because of, its symbolic value, the Arabian oryx was often hunted in Arabia. Traditionally, this was partly due to a local conviction that the horns give sexual vigour when ground into powder and eaten.[31]

In the middle of the twentieth century, hunters drove the Arabian oryx to the brink of extinction. In the 1940s, oryx were still seen on a regular basis. An American geologist surveying Saudi Arabia once saw two herds in a matter of days. However, in the 1950s, hunters began to use automatic weapons and cars followed by helicopters. Spotted by aircraft, chased by land rovers and exposed in the open desert, the oryx hardly stood a chance.[32] One oryx population in the Nafud desert of northern Arabia was hunted to extinction during the 1950s. A second population lived in and around the southern Rub' al-Khali, distant from most human settlements. However, the hunting pressure increased even on this population during the 1950s and 1960s, causing international conservationists to have little hope for the survival of the Arabian oryx in the wild.[33] Hunting also created the perception that Arabia may be the 'natural environment' of the species, but not the best place for its conservation.

At this point, a network of international conservationists ventured to save the Arabian oryx by taking some members of the species out of its habitat. In 1962, Ian Grimwood, the chief game warden of Kenya, led 'Operation Oryx' in the British protectorate of Aden. Organised by the

Fauna Preservation Society, this operation expedition received financial help from the recently established World Wildlife Fund.[34] Just inside the borders of Oman, Grimwood and his companions sighted four oryx that had survived a recent encounter with a prince's hunting party.[35] Grimwood's team lassoed three of them from a specially built car and captured them alive. Subsequently, the Fauna Preservation Society planned the establishment of a 'world herd' that would be bred in captivity. After a long investigation, Grimwood decided that the best place for this world herd was not Kenya, but Arizona, whose climate resembled that of Arabia. The three animals from Arabia and a fourth one donated by the Zoological Society of London (ZSL) were subsequently shipped to Phoenix Zoo.[36]

The removal of Arabian oryx from the British protectorate of Aden could be interpreted as a late appropriation of a colony's natural resources. Yet, the establishment of the world herd also continued older patterns of animal gift exchange between aristocrats, some of whom had come to head international conservation organisations. During the decades after World War II, the Gulf rulers became increasingly concerned about the possible extinction of the Arabian oryx. As a part of 'Operation Oryx', conservationists asked a few of them to donate their animals to the world herd. Sheikh Jabir Abdullah Al Sabah of Kuwait offered two animals to the Fauna Preservation Society, one of which survived and joined the herd at Phoenix Zoo. After several approaches, Grimwood also negotiated a gift of four animals by King Saud bin Abdulaziz of Saudi Arabia to the WWF, which was headed by Prince Bernhard of the Netherlands. Grimwood, who exchanged the four oryx for some animals from Kenya,[37] considered them 'a valuable second string to the Phoenix breeding centre'. In 1964, he collected these animals in Riyadh, where they were flown first to Naples and then to New York, before joining the world herd in Phoenix.[38] Unlike the 1979 flight, this transfer via Italy went without major complications.

Besides aristocratic links, connections with oil companies were important in bringing Arabian oryx to Arizona. In 1965, Grimwood noted that British and American oil companies in Arabia shared information and provided 'practical help wherever possible'. The Arabian American Oil Company had 'built the crates and carried out all arrangements for the removal of the oryx from Riyadh'. The WWF sponsored the further transport of the oryx, but received 'financial help' from Aramco.[39] The support by oil companies for conservation perhaps arose from the potential of conservation in public relations and the personal interests of oilmen in wildlife. The company magazine *Aramco World* frequently featured articles on Arabian wildlife conservation.[40]

While the princes and oil companies in Saudi Arabia and Kuwait contributed to the formation of a world herd in Arizona, the Sultanate of Oman sought to conserve the oryx inside its borders. In the early 1960s, Oman had probably the largest population of wild oryx in the Arabian peninsula, estimated at around 100. In order to protect them, Sultan Said bin Taimur issued a decree that banned oryx hunting from vehicles. His successor Qaboos renewed this decree and gave special instructions to the Sheikh of the local Harasis tribe to protect the oryx. The local Bedouin mostly respected this ban. Nevertheless, well-equipped and organised foreign hunting parties continued to enter Oman from the north.[41] In 1972, poachers killed the last wild oryx in Jiddat al-Harasis in the central desert of Oman.[42]

Subsequently, Sultan Qaboos supported the reintroduction of the oryx in the context of wider governmental efforts to map and conserve Oman's natural resources and heritage. People with backgrounds in colonial administration and oil companies were again central to these efforts. An important figure in Omani policies was the British naturalist Ralph Daly, who had developed an interest in Arabian wildlife while serving in Aden. After his departure from Aden, he worked in the Government Relations Department of Petroleum Development (Oman). Through this work, Daly's interest in local flora and fauna came to the attention of Qaboos. In 1974, the sultan appointed Daly as an advisor on the conservation of the environment.[43] After agreeing on the conservation of the Arabian tahr in the mountains of northern Oman, Qaboos ordered Daly to examine possibilities to restore the oryx as part of the country's 'national heritage'.[44]

In order to reintroduce the oryx, the Omani government reached out to the conservation organisations that controlled the world herd in Arizona. In 1975, Daly helped Oman become a member of the International Union for Conservation of Nature, and took an Omani delegation for the first time to the IUCN General Assembly.[45] 'At that stage', Daly later said, 'the World Wildlife Fund had hardly even heard of Oman.'[46] However, Daly managed to arrange a visit by Hartmut Jungius from the IUCN and the WWF to study the feasibility of a reintroduction of the oryx to Oman.[47] Together with other scientists, Daly and Jungius travelled extensively through the interior of Oman in 1977 and 1978. They found that the best location for the reintroduction was Jiddat al-Harasis, the plateau of the Harasis tribe. 'Ironically, yet logically', this was the place where the last oryx had been killed only five years earlier.[48] An unfenced Arabian Oryx Sanctuary the size of Belgium or Massachusetts[49] was created, offering the oryx a lot of vegetation.

Another network exchanging money, environmental credentials, expertise and animals thus emerged between aristocrats and conserva-tionists. Already in 1974, Sultan Qaboos realised that reintroducing the oryx would require 'careful planning' and 'a lot a money'.[50] Beginning in 1979, animals from the world herd in Phoenix Zoo were transferred to Jiddat al-Harasis. Hartmut Jungius recommended that the world herd should provide 12 oryx, which would be acclimatised before being released in Jiddat al-Harasis. The trustees of the world herd, which included the WWF, the Fauna Preservation Society and the Zoological Society of London, agreed to this plan, and the animals were subsequently shipped to Oman.[51] This reintroduction received, and gave aristocrats and conservationists, much publicity. Qaboos flew into the area to inspect the project in front of the media. In 1982, Prince Philip, the Duke of Edinburgh and president of WWF International, came to see the oryx in the wild.[52] Daly continued to work closely with the IUCN and the WWF and, in 1990, secured the sultanate's 'generous support' of the IUCN's Sir Peter Scott Fund for on-the-ground conservation projects.[53]

However, the Arabian oryx reintroduction project did not remain at the level of international organisations and politicians. For the execution of the project, support from a local tribe was as important as late colonial and postcolonial conservation connections. In order to protect the oryx, the Omani government hired men from the local Harasis tribe as rangers. Daly supported this recruitment as a way of allowing the Harasis to continue their 'traditional pursuits and prevent them from becoming the low-paid labour of the towns and oil camps'.[54] For the Harasis, who numbered about 350 in 1982, this was not only a job and a form of patronage by the sultan, but also a matter of tribal honour. They had been outraged when hunters drove into their territory in 1972 and wiped out the last herd. They thought of the oryx as their tribal property and pledged the support of their entire tribe for the reintroduction project.[55] At the same time, they gained an opportunity to display their loyalty to their patron. When the first oryx returned to Oman, the old men thanked Qaboos, and the young men even cried, 'So that is what they really look like; our elders did not lie. Praise be to Allah and the Sultan.'[56]

Thanks to the collaboration between the conservationists and tribal rangers, the oryx reproduced in Jiddat al-Harasis during the 1980s and early 1990s. In 1982, one of the females went into labour and produced a live calf. It was the first to be conceived in Oman and born in the wild in ten years. The birth was also remarkable as the same female had borne a calf in an enclosure earlier and had refused to nurse it. This time she took care of her newborn instantly, and the father began to stand guard over mother and calf. In 1990, the population in the Arabian Oryx Sanctuary passed 100.[57] In 1994, the United Nations Educational, Scientific and Cultural Organization (UNESCO) named the reserve – as the first natural place in the Gulf monarchies[58] – a World Heritage Site.[59] The connections with international experts and organisations thus also enabled the sultanate to appear as a 'regional leader in biodiversity conservation'.[60]

The conservation efforts not only resulted in oryx offspring, but also in scientific publications. Already in the mid-1970s, Ralph Daly realised the need for research into desert ecology. He assumed that the oryx had disappeared because of overhunting, but also considered possible other reasons, such as habitat destruction and climate change. He was thus interested in research on the ecological requirements of the Arabian oryx and the natural history and human use of the reintroduction area.[61] As manager of the reintroduction project, Daly hired a British researcher named Mark Stanley-Price. Previously, Stanley-Price had spent six years in Africa attempting to domesticate the African oryx, a cousin of the Arabian oryx. Because the oryx were immune to the tsetse fly, he had sought to develop them as livestock. In Oman, Daly charged Stanley-Price with the opposite. He tried to wean the Arabian oryx who came from the Phoenix Zoo from the habits of domestication and trained them to live in the wild.[62]

Stanley-Price used the evolutionary concept of adaptation in order to explain what enabled oryx to survive in the wild and what gave them their iconic morphology. After spending over seven years in Oman during the 1980s, the conservationist published a book about his work.

In contrast to the book *Oases of Oman*, Stanley-Price's book appeared outside of Oman. Hence, it did not undergo Omani censorship, but probably also reached fewer people in Oman. The British conservationist explained that Arabian oryx calves are sandy brown for the first two to three months of their lives, which is 'an obvious adaptation to reducing predation risk'. As they grow up, oryx become white to minimise the radiant heat from the sun and to locate one another more easily at a distance. Stanley-Price also stated that 'the more desert-adapted Arabian and scimitar-horned oryx have evolved their conspicuousness as adults' due to the absence of many predators in their habitat.[63]

During the 1990s, the oryx programme in Oman ran into difficulties. One of the main reasons was that the government's patronage network, which sustained the local rangers, ignored a socioeconomic divide between two tribes. The government trained and employed guards, rangers and guides only from the Harasis tribe in central Oman. While this made sense because of the Harasis' familiarity with the area, it was also part of a wider pattern in the Gulf monarchies to fill some governmental positions almost exclusively with members of certain families, clans and tribes. In the case of the Arabian Oryx Sanctuary, the hiring of the Harasis meant an exclusion of the large Janabah tribe to the east, which mostly consisted of relatively poor pastoralists and fishermen. Animal collectors, mostly from the Gulf region, exploited the rising socioeconomic divide between the Harasis and the Janabah. From 1996 onward, they offered Janabah cash to poach oryx, as much as $25,000 for a live female. The eight Harasis rangers who were charged with patrolling the huge, unfenced reserve became unable to protect the oryx.[64]

The patronage network of the Omani government thus failed as a safety net for the Arabian oryx. Because the reintroduction had been so successful, it attracted the interest of an illegal regional trade in wildlife and resulted in a resurgence in poaching.[65] Within a decade, the oryx herds were decimated. Some animals died from injuries or stress during attempted captures. The rangers found other oryx dead and dumped in the desert by fleeing poachers. Many others were taken and sold alive, mostly to private menageries. Between 1996 and 2007, the number of free-roaming oryx thus declined from 450 to 65, only four of them females.[66] Deeply saddened by this development, Ralph Daly died in 2006.[67] In 2007, a decree by Sultan Qaboos reduced the Arabian Oryx

Sanctuary by 90 per cent to a size that could be effectively guarded.[68] UNESCO feared that the reduction in the size of the reserve and plans to proceed with hydrocarbon prospecting would destroy the value and integrity of the property.[69] It thus delisted the sanctuary as a World Heritage Site, the first time that this had been done. Virtually all of Oman's remaining oryx were gathered into a fenced enclosure of four square kilometres, where they were protected by police patrols with automatic weapons. A government advisor commented, 'The idea of free-ranging wild herds is finished.'[70]

## Animal kingdoms

While conservation in Oman experienced difficulties and setbacks, other Gulf monarchies also sought to protect and reintroduce Arabian oryx and other large mammals. In 1969, Zayed bin Sultan, the ruler of Abu Dhabi, presented three oryx, which were confiscated from a hunter, to Al Ain Zoo. By 1982, the herd at this zoo numbered 22 animals. In the same year, Bahrain's Al-Areen Wildlife Park housed not only the antelopes rescued from the plane in Rome in 1979, but also 14 oryx.[71] In order to protect them from hunting, Al-Areen Wildlife Park started an education programme. 'We hope', the park's director said, 'to implant a conservation ethic in young people that will be passed on to future generations.' The park's administrative manager added, 'We're trying to show the role man plays in maintaining the balance of nature.' Between 1983 and 1984 alone, 5,000 of Bahrain's schoolchildren were shown around the park.[72]

While all the countries on the Gulf Cooperation Council engaged in similar conservation work, Saudi Arabia, the largest of them, invested the most in conservation and associated research. This investment involved a triangular connection between the Saudi government, scientists and business. During the 1970s, zoologists working for foreign companies and the Saudi government started producing comprehensive surveys of the Saudi animal kingdom. An important individual was the Swiss entomologist Wilhelm Büttiker, who came to Saudi Arabia in 1975 to work for Ciba-Geigy on a public hygiene project. Ciba-Geigy, which was based in Basel, had a long-standing interest in research on insects. One of the company's researchers had discovered the quality of DDT as an insecticide in 1941. Subsequently, the company became a major producer of DDT, which was widely used to defeat malaria and defend crops in the Middle East.[73] Ciba-Geigy's project in Saudi Arabia, which the company carried out with the Ministry of Municipal and Rural Affairs, allowed Büttiker to collect insects from various regions of the kingdom. Realising that no comprehensive research on the kingdom's fauna existed, Büttiker initiated, outside his normal duties, the Zoological Survey of Saudi Arabia.[74] In 1978, Ciba-Geigy consented to Büttiker's full-time employment as a research zoologist and provided laboratory and transport facilities for his work.[75]

In order to benefit from international expertise, Büttiker established a network with scientists in Europe. 'During this period', the Swiss entomologist stated, 'Faculties of Science were founded at several universities in Saudi Arabia and neighbouring countries. The fieldwork in conducting the Zoological Survey was a novelty to the region's academic curricula. It needed mental resolve, time and funding.'[76] In the absence of experienced Saudis, Büttiker's main collaborator was a fellow Swiss entomologist named Walter Wittmer from Naturhistorisches Museum Basel. Wittmer joined Büttiker in two expeditions to the kingdom in 1976 and 1978. In order to publish their results, the two entomologists founded the series *Fauna of Saudi Arabia* in 1979, which subsequently became one of the leading platforms of biological research in the Gulf. Ciba-Geigy and the fund Pro Entomologia of Naturhistorisches Museum Basel jointly published the first two volumes.[77]

The connections with Europe were also important for the replacement of scientists who withdrew from the Swiss zoological network. After the publication of volume five in 1983, Wittmer retired from his position at Naturhistorisches Museum and as co-editor of *Fauna of Saudi Arabia*. Büttiker searched for successor and found a German ichthyologist named Friedhelm Krupp,[78] who worked as a research fellow for Tübinger Atlas des Vorderen Orients. Krupp had 'dreamed' about going to Saudi Arabia, but found it 'impossible to gain a visa'. He thus applied to Büttiker to work on the fishes of Saudi Arabia and met him in Basel. Soon thereafter, in 1984, Krupp joined Büttiker as co-editor of *Fauna of Saudi Arabia*.[79]

The European editors of *Fauna of Saudi Arabia* served as hubs between scientists inside and outside the Gulf. While Büttiker and a few other scientists collected specimens during their expeditions inside the kingdom, specialists based elsewhere wrote most of the articles. In order to sustain this network, Büttiker's connection to Wittmer at Naturhistorisches Museum Basel was pivotal. Büttiker sent materials from Saudi Arabia to Wittmer, who processed and despatched them to colleagues around the world. This also saved these colleagues from the difficulties of obtaining visas and research permits for Saudi Arabia. By 1981, over 100 scientists had received material, including David Harrison, an authority on Arabian mammals who led his own Harrison Institute in England. Other scientists were specialists on insects and reptiles based at institutions, such as London's Natural History Museum,[80] which were themselves colonial and postcolonial centres of calculation. The travel of specimens and articles along these global networks allowed for the fast identification and description of many new Arabian animals. By 1990, *Fauna of Saudi Arabia* had 200 contributors from around the world, who had described almost 800 new genera, species and subspecies. Fifty-four of these taxa were named after Büttiker,[81] recognising the central role of the Swiss entomologist in the Gulf's zoological networks.

While few Saudi scientists figured among the authors, Saudi institutions became integral parts of the networks of zoological research. In 1981, the MEPA acquired and started sponsoring *Fauna of Saudi Arabia*. In the same year, Büttiker retired from Ciba-Geigy and started working as a consultant for MEPA.[82] Although Ciba-Geigy thus withdrew from the series, the editors still acknowledged a 'generous donation' by the company in 1985. This donation reflected 'continuing interest' by the company's chief executive in zoological research in the kingdom.[83] Abdelelah Banaja, a Saudi zoologist working for MEPA, justified the administration's sponsorship of the series in economic as well as environmental terms, implicitly challenging the primacy of oil in the national economy. 'One of the main tasks of MEPA is to consolidate a firm basis of environmental knowledge to which this series, comprising a comprehensive survey of animal life in Saudi Arabia, may well contribute,' Banaja wrote. 'It will serve as the scientific data base for defining the steps necessary for appropriate preservation of our most invaluable resource: Nature.'[84]

Independent from the work of Büttiker and the MEPA, members of the Saudi royal family cultivated their own connections with European conservationists. King Khalid, a passionate hunter, initiated a breeding programme for Arabian oryx in Saudi Arabia. This seems paradoxical, but hunters, such as Prince Charles, the president of WWF United Kingdom, were often among the founders and patrons of environmental and conservation movements in various national contexts.[85] Khalid concentrated some oryx from Saudi Arabia, neighbouring countries and the world herd at his farm at Thumamah, north of Riyadh. However, due to overcrowding in the enclosure, diseases spread among the animals, and calf mortality was high.[86]

**Figure 3.1** Saud Al-Faisal at the entrance of the National Wildlife Research Center (courtesy of Stéphane Ostrowski and the NWRC)

After King Khalid's death in 1982, other Saudi princes became more involved in conservation. The most prominent was Saud Al-Faisal, the foreign minister and a son of Khalid's predecessor Faisal. A self-proclaimed 'reformed hunter', Prince Saud was particularly concerned about the disappearance of the houbara bustard, the most important prey in falconry.[87] As in the case of the Arabian oryx, hunting and habitat loss had reduced houbara numbers dramatically during the twentieth century. Prince Saud thus consulted with Jacques Renaud, a veterinarian and entrepreneur, who had established commercial wildlife parks in France during the 1960s.[88] An expert in falconry and bird breeding, Renaud had joined the Saudi Crown Prince Abdullah bin Abdulaziz and other princes during hunting trips in the desert. In consultation with Saud Al-Faisal, the French veterinarian prepared a proposal for a houbara breeding and reintroduction project that would be managed by Reneco, a company owned by Renaud.[89]

Besides being a profitable venture for a French natural entrepreneur, the houbara project became the core of a new Saudi government agency that offered scientists patronage and support. After receiving Renaud's proposal, Saud Al-Faisal invited the president of the Saudi Biological Society Abdulaziz Abuzinada to develop it. A botanist and not a falconer, Abuzinada had not heard of the houbara by that time. He agreed with Renaud's houbara project 'in principle', but considered the focus on one species alone too narrow and insisted that conservation cover the entire biodiversity of the kingdom.[90] Prince Saud agreed, and Abuzinada in collaboration with others conceived of a new organisation, which he called the National Commission for Wildlife Conservation and Development (NCWCD).[91] (The amateur botanist and collaborator of the commission Sheila Collenette called it 'a huge mouthful'.)[92] In 1986, King Khalid's successor Fahd formally established this commission by royal decree. One of its aims was to 'develop and implement plans to preserve wildlife in its natural ecology and to propose the establishment of proper protected areas and reserves for wildlife in the Kingdom'.[93]

Because of the interests of senior princes in hunting and conservation, leading members of the Saudi government and the scientific profession supervised the National Commission for Wildlife Conservation and Development. Despite small budgets compared with most other ministries, these people provided the commission with protection from the rest of the bureaucracy. This was similar to the Royal Commission for Jubail and Yanbu, which acted as a support agency for the Saudi Basic Industries Corporation (SABIC).[94] In 1992, the budget of the NCWCD totalled $17.2 million.[95] Saud Al-Faisal was the managing director of the Commission and Abuzinada the secretary general. The chair was Sultan bin Abdulaziz, the minister of defence and aviation and second deputy prime minister. Other members of the board included the minister of agriculture and water, the president of the King Abdulaziz City of Science and Technology, and the president of the MEPA.[96]

The appointment of such a powerful board perhaps sought to strengthen the NCWCD, especially in dealing with MEPA and the MAW. Initially, both MEPA and the NCWCD claimed the Asir National Park as under their control.[97] Typical of the lack of horizontal integration between Saudi government agencies,[98] there was also little administrative communication between MEPA, the NCWCD and MAW. The MAW generally implemented agricultural projects without the supervision of either MEPA or the NCWCD. On few occasions, ad-hoc committees comprising representatives from different agencies discussed specific cases. Finally, in the absence of environmentalist political parties and strong pressure groups within civil society, the NCWCD needed members of the ruling family and ministers to be more effective.[99]

Backed by senior princes, like Saud Al-Faisal and Sultan bin Abdulaziz, the National Commission for Wildlife Conservation and Development took over responsibilities from the MEPA. In his preface to volume nine of *Fauna of Saudi Arabia* in 1988, Prince Saud announced the transfer of the series from MEPA to the NCWCD. This was part of the commission's plan to 'encourage and conduct scientific research on wildlife biology and ecology'.[100] Like MEPA officials, Saud Al-Faisal justified his support by referring to the notion of 'sustainability', implicitly challenging the primacy of the oil sector. Coming from a senior prince, this commitment to a diversification of the economy was probably uncontroversial. However, it showed an increasing perception of Arabia's living nature – as opposed to its dead fossil fuels – as economically valuable. In another preface in 1993, Prince Saud stated that *Fauna of Saudi Arabia* 'is of great value in promoting further scientific research and in forming the basis for the applied fields. It thereby plays a role in the development of a sustainable economy.'[101]

In order to conserve houbaras and other animals, the National Commission for Wildlife Conservation and Development re-activated the older system of *ḥimá*s or 'sanctuaries'. Besides sponsoring the series *Fauna of Saudi Arabia*, the commission established protected areas for wildlife in order to restore and preserve the kingdom's biodiversity. Ultimately, the reserves were also supposed to 'rehabilitate game to levels where it can again be hunted on a sustainable basis'. By 1995, the commission had established 15 protected areas with a total size of around 90,000 square kilometres.[102] In a continuation of previous conservation practices, these areas encompassed some of the older tribal *ḥimá*s.[103] The NCWCD sought to conserve different species of mammals and birds as well as reintroducing others, like the Arabian oryx, to these areas.[104]

In order to support its conservation and reintroduction efforts, the National Commission for Wildlife Conservation and Development established breeding and research facilities near the protected areas in addition to a headquarters in Riyadh. In 1986, the commission founded the National Wildlife Research Center (NWRC) near Ta'if. The NWRC was close to the protected area Mahazat as-Sayd, whose name indicated the previous use of the area for 'hunting' (*ṣayd*). Yet, unlike the large Arabian Oryx Sanctuary in Oman, Mahazat as-Sayd was completely fenced in as protection against hunters and livestock grazing. The primary aims of the NWRC were the captive breeding of houbara bustards and Arabian oryx for eventual reintroduction into Mahazat as-Sayd and other areas. The NCWCD thus decided to relocate the oryx herd from the late King Khalid's farm to the NWRC. In 1986, 57 oryx flew to Ta'if in communal crates, assisted by the staff of the Zoological Society of London.[105]

Like Wilhelm Büttiker with his Zoological Survey of Saudi Arabia, the NWRC subsequently formed a hub connecting princes with scientists and business. At the same time, the centre enjoyed autonomy from the rest of the Saudi bureaucracy. Under the protection of princes and with the support of the NCWCD, the centre thus formed an 'island of efficiency' among state-owned organisations, similar to SABIC.[106] Jacques Renaud's company Reneco ran the NWRC for profit, while Abuzinada charged one of his clients, the Saudi botanist Abdul-Rahman Khoja, with coordinating the activities of the centre with the NCWCD's headquarters in Riyadh.[107] Following the plans of Jacques Renaud, who served as consultant manager, French and other Western scientists formed most of the centre's administrative and scientific staff.[108]

Partly in order to generate income for Reneco, the NWRC actively sought new research projects in order to maintain or increase its budget. In 1986, four veterinarians, one ecologist, two specialised breeders and support staff were working at the centre.[109] 'Much of the science that was accomplished at NWRC was done when the centre was under the Auspices of Mr Jack Renaud', said the American ornithologist Joe Williams. 'Without the French, very little would have been accomplished', he added. Besides Renaud, another Frenchman named Patrick Paillat 'was instrumental in encouraging all kinds of research projects at the centre', Williams claimed, 'even though he did not understand them'.[110] While this encouragement sometimes distracted from the centre's core objective of houbara breeding, it also provided freedom for the pursuit of a range of research interests.

Soon after the establishment of the NWRC, the NCWCD transformed the late monarch's farm near Riyadh into the King Khalid Wildlife Research Center (KKWRC). In contrast to the NWRC, which focused on houbara bustards and Arabian oryx, this second research

**Figure 3.2**   National Wildlife Research Center (courtesy of Stéphane Ostrowski and the NWRC)

centre concentrated on gazelles. In 1987, the NCWCD founded the KKWRC and gave the responsibility for its management to the Zoological Society of London in October.[111] These contracts with organisations from different countries (France in the case of the NWRC, and Britain in the case of the KKWRC) were a common Saudi way of balancing different foreign influences.[112] The ZLS conceived the KKWRC as a centre that employed up to 40 international scientific and support staff. On behalf of the NCWCD, the ZSL advertised the positions of project manager and station manager in the British magazine *New Scientist* in 1988. Tax-free inside Saudi Arabia and combined with other allowances, the salaries were competitive. For up to £30,000 for the project manager and up to £18,000 for the station manager, the ZLS asked candidates to 'enjoy challenging conditions'.[113]

The conditions at the new wildlife research centres were challenging, indeed. As with other projects of rentier science that relied on personal connections, the death of King Khalid as their patron had proven devastating. Five years after he had died in 1982, his farm had severely deteriorated, and tuberculosis had spread among the animals. This epidemic became apparent in 1986, when the 57 oryx were flown from Riyadh to Ta'if. The flight was so stressful that three animals died of trauma and one female aborted.[114] Two months after the flight, in June 1986, one oryx suddenly died of tuberculosis. By September 1987, 16 animals had died of the disease.[115] Consequently, the NWRC directed much of its activities at controlling the disease between 1986 and 1988.[116]

Despite these difficulties, the symbolic importance of the oryx meant that the scientists at the NWRC continued to receive support from the leadership of the NCWCD. Personal patronage thus allowed for a long treatment that would have been difficult to fund outside the Gulf's rentier systems. Looking back at their management of the disease, French scientists stated that 'the elimination of tuberculosis from this herd of Arabian oryx was very costly and could be justified only by the rarity of the specimens and their genetic value'. The scientists used expensive medication imported from Europe and isolated calves at birth in order to create a tuberculosis-free generation of oryx. The scientists thanked Saud Al-Faisal and Abdulaziz Abuzinada personally 'for their support, confidence and patience throughout this long-term project.'[117]

After having controlled the tuberculosis outbreak, the researchers at the NWRC succeeded in breeding a healthy population of Arabian oryx. In a public reintroduction ceremony, this population then served its patrons as natural resources in the symbolic economy. In March 1990, Sultan bin Abdulaziz, the chair of the NCWCD, released the first group of 17 animals into the Mahazat as-Sayd Protected Area. These releases publicly linked the princes with conservation, winning favourable headlines and environmental credentials. Animals from neighbouring countries subsequently reinforced the oryx released by Prince Sultan. By 1995, a third generation of calves had been born in the protected area, bringing the population to approximately 180.[118]

Thanks to the connections with the senior princes, zoological research and conservation in Saudi Arabia continued during periods of lower oil prices and relative fiscal austerity during the late 1980s and 1990s. This was in contrast to the botanical research by the Germans Harald Kürschner and Wolfgang Frey that did not enjoy royal patronage. In 1989, Wilhelm Büttiker and Friedhelm Krupp wrote in their introduction to volume ten of *Fauna of Saudi Arabia* that 'despite the current financial situation worldwide, budgetary constraints have so far had little impact on the format and scholarly standards of the series'. Büttiker and Krupp also conveyed 'their heartfelt thanks and appreciation to the National Commission for Wildlife Conservation and Development of the Kingdom of Saudi Arabia for the enthusiastic support that has enabled this series to continue'.[119] In 1998, oil prices reached another low because of the Asian financial crisis. In this situation, senior members of the royal family gave further private funds to the commission. In June 1998, Sultan bin Abdulaziz announced donations of 10 million riyals (around $3 million) by King Fahd and 5 million riyals by Crown Prince Abdullah bin Abdulaziz to the NCWCD. By reporting on these donations, the Saudi newspaper *al-Jazirah* further added to the environmental credentials of the royal family.[120]

While the princes provided patronage for the foreign scientists, the top-down decision-making in rentier science and politics also disappointed some researchers. The relationship between the princes and scientists thus also experienced tensions, and the exchange of money, expertise, animals and environmental credentials was not entirely secure. A British conservationist, who worked at the KKWRC between 1987 and 1990, expressed some frustrations. He acknowledged that the

'National Commission for Wildlife Conservation and Development has done a great deal in the first 5 years of its existence, and Saudi Arabia probably makes a greater financial commitment to conservation than any other Arab country.' The conservationist also lauded some of the expatriate experts for having 'considerable experience in wildlife conservation and management'. Yet, he lamented that these experts were 'not used effectively, because none is in executive positions (except in the breeding centres) and their advice is often ignored'. He added that the commission's 'board of directors consists largely of senior members of the royal family who make most major decisions without consulting the scientific staff'. Finally, according to the British conservationist, the continued participation by these princes in 'destructive forms of hunting' seriously damaged the 'international image' of the commission.[121]

## Publishing evolution

The lack of consultation between scientists and senior princes was an important *dis*connection in the history of biology in the Gulf. Paradoxically, this disconnection also gave the researchers considerable freedom and space, which allowed them to work on sensitive topics, such as evolution. It thus also created a certain mismatch between the creationism of many officials and the evolutionary terminology of many scientists. This creationism appeared in the publications of the MEPA, which supported the Zoological Survey of Saudi Arabia and propagated an Islamic environmentalism in the 1980s. In 1985, Abdulbar Al-Gain, the deputy director of MEPA, wrote a preface to volume seven of *Fauna of Saudi Arabia*, which he began with a Qur'anic verse: 'Verily, all things have We created by measure.'[122] Sultan bin Abdulaziz, the NCWCD's chair, was also a creationist, even though Abdulaziz Abuzinada, the secretary general, was a proponent of evolution. In 1994, Prince Sultan wrote that 'conscious of its responsibility towards God's creation, the Government of the Kingdom of Saudi Arabia has always given the highest priority to the protection' of the plant and animal life of the Gulf region.[123]

Despite the disconnection between certain scientists and princes, there was no absolute divide between scientists on the one hand and officials on the other in terms of adherence to creationism or evolutionism. Sultan bin Abdulaziz believed in adaptations alongside creation, writing that the Gulf's flora and fauna had 'adapted in a most remarkable way to the harsh environment'.[124] Some of the foreign scientists also shared the creationism of princes and officials, resulting in creationists and evolutionists working side-by-side in many research institutions. A pertinent example is the entomologist Abdul Mon'im Talhouk who worked together with the evolutionist Shaukat Chaudhary for the Saudi Ministry of Agriculture and Water.[125] During the 1970s and 1980s, he collaborated with Wilhelm Büttiker and contributed important collections of insects to the Zoological Survey of Saudi Arabia.[126]

Through his work for the Zoological Survey of Saudi Arabia, Talhouk was able to propagate creationism and religious views of nature in the series *Fauna of Saudi Arabia*. Saudi-sponsored publications thus did not separate science and religion, even if Europeans edited them. In his introduction to volume two of *Fauna of Saudi Arabia* in 1980, Talhouk wrote, 'In His wisdom, God has created all creatures, including man, to interact. The sacred verses that crown His introduction are a witness to His love and special care for man's welfare.' Talhouk also shared the view of scientific exegesis that the Qur'an contains modern scientific knowledge. He stated, 'As science keeps advancing, man will discover more of the hidden wisdom in God's Book.' Continuing his introduction, Talhouk justified conservation religiously, echoing MEPA's Islamic environmentalism. He stated that 'it behooves us, to remember our duties, as true believers, to protect His great creations, creations that He entrusted to us as custodians.' Like Sultan bin Abdulaziz, Talhouk combined creation with adaptation, writing that Saudi Arabia's species had 'interacted for thousands of centuries, adapting themselves to suitable habitats'. Finally, he also warned against the dangers posed by agriculture and development. He stressed that 'preservation of these highly adapted ecosystems is urgently needed', with natural ranges being threatened by 'overgrazing and the pressure of civilisation'.[127]

Despite creationist and adaptionist prefaces and introductions, the editors of the series *Fauna of Saudi Arabia* accepted articles that referred explicitly to evolution. Hence, there was no strict division between creationist and evolutionist publications in the Gulf. Volume ten of *Fauna of Saudi Arabia* contained an article entitled 'Evolutionary History and Zoogeography of the Red Sea Ichthyofauna'. Its author was the ichthyologist Wolfgang Klausewitz, who worked for the Senckenberg

Research Institute and Natural History Museum in Frankfurt, the same institution that employed the co-editor of *Fauna of Saudi Arabia*, Friedhelm Krupp. Klausewitz wrote that 'the evolution of endemics among the secondary deep-sea fishes started during the Pleistocene, probably during the second or third interglacial period, as the process of speciation generally needs more than 10,000 or 15,000 years'.[128]

There is no sign of any Saudi governmental agency censoring Klausewitz's article and trying to remove evolution. The National Commission for Wildlife Conservation and Development co-published the volume in which the article appeared, together with the Swiss fund Pro Entomologia. Yet, rather than passing through a Saudi agency, the manuscript went through the Western network of peer reviewers. In his acknowledgements, Klausewitz thanked three researchers for 'reviewing and substantially improving the manuscript'. One of them was *Fauna of Saudi Arabia*'s co-editor Friedhelm Krupp, with the other two being ichthyologists at the University of Florida and the National Museum of Natural History in Washington.[129]

Klausewitz's article also indicates that there was little religious censorship of scientific publications, such as *Fauna of Saudi Arabia*, in the Gulf. Friedhelm Krupp even claimed that there was no repression of the theory of evolution among academics in the Gulf. He stated that during his decades as co-editor, evolution 'had not been an issue'. 'Occasionally, we mentioned it to our colleagues', he stated. A common response was, 'Of course, we teach evolution, and that makes perfect sense.' Krupp also compared the situation in the Gulf monarchies with that in another oil state, Texas. The director of the Houston Museum of Natural Science had once told Krupp that he had been 'forbidden to use the word evolution anywhere in our exhibition'. Krupp commented, 'something like that never occurred to me in Arabia'.[130]

One might argue that the series *Fauna of Saudi Arabia* was produced largely by and for foreign specialists and was thus free to include research into evolution. However, the theory of evolution also entered publications aimed at non-specialist audiences. The National Commission for Wildlife Conservation and Development sponsored such publications as part of wider efforts to create public awareness of the protection of nature, efforts that included the production of posters, television programmes and press releases.[131] Such a publication for a wider audience was the English-language magazine *Arabian Wildlife*, which the NCWCD launched in print in 1994 and later published online. The magazine contained mainly short articles written in simple language and lacking notes or bibliographies.

While raising conservation awareness, publications such as *Arabian Wildlife* allowed foreign scientists to communicate with non-specialists, occasionally covering evolution. One of these scientists was the rodent specialist Paul Bates, a collaborator of David Harrison. In 1984, Harrison and Bates published an article on bats in *Fauna of Saudi Arabia*.[132] Later, the two zoologists worked on a second edition of Harrison's book *Mammals of Arabia*. In turn, Harrison provided Bates with 'most useful comments', when the latter published an article on desert mice in *Arabian Wildlife* in 1995. Although Bates was addressing a wider audience, evolution was central to his narrative. The zoologist stated that 'the evolutionary prototypes of the present-day jerboas probably originated some 8 million years ago on the arid plains of Central Asia'. Later in the article, Bates described the rare Arabian garden dormouse as 'an evolutionary off-shoot of the tree-dwelling European dormouse', which had 'become adapted to a non-arboreal existence during the last 1.2 million years'.[133]

Many articles in *Arabian Wildlife* were not primarily concerned with evolution, but some of them still expressed idioms and convictions associated with it. Hence, the Saudi government did not effectively censor even a non-specialist publication such as *Arabian Wildlife* as regards to scientific theories. This is evident in an article by was Jacques Flamand, a French scientist involved in oyrx conservation at the National Wildlife Research Center. Flamand wrote that after their release into the Mahazat as-Sayd Protected Area, the oryx had successfully 'adapted to their new environment'. He reported that some males had been killed in combats with other males, explaining that 'these losses must be considered as part of the normal life of wild oryx with competition ensuring that the fittest survive and helping to maintain the herd's overall health'.[134]

In many instances, Saudi governmental agencies did not publish research into evolution themselves, but supported research that ultimately appeared in non-Saudi journals. This was the case with primatology, a small field compared to studies on ungulates. In contrast to Arabian oryx, the Saudi princes showed comparatively little interest in the hamadryas baboons, the only extant non-human primate species in Saudi Arabia. In the animal economy of agriculture, gift exchange, conservation and research, primates played a less important role than camels or antelopes. The ninth-century Abbasid caliph al-Mutawakkil once received two monkeys from a king of Nubia.[135] However, as baboons remained relatively abundant, they were of less value to monarchs and conservationists. Hans Kummer, a Swiss primatologist, offered a further explanation for the scarcity of researchers on hamadryas, although the ancient Egyptians had worshipped them as sacred animals. Most mammals entice each other to mate using scents imperceptible to humans, or sounds that humans do not recognise as sexual. However, 'baboons are visual animals, and their physical displays and sexual exuberance are understandable to humans as such', said Kummer. 'What's worse', baboons are 'naked from the waist down.' To some, they 'are simply embarrassing'.[136]

The lack of research interest in baboons probably also related to their ambivalence in Muslim cultures. On the one hand, Muslim natural philosophers placed primates over other animals and close to humans in the 'chain of being' based on Aristotelianism.[137] On the other hand, likening a man to a monkey, in contrast to lions and other wild animals, was a degrading insult.[138] Baboons were almost absent from the rich Arabic poetic tradition compared to oryx, gazelles, wolves and even hyenas. 'Baboons are neither beautiful to describe, nor good to eat', said Ahmed Boug, a Saudi primatologist and poet, 'So the poets ignored them.' Boug added that some people in south-west Arabia believed that 'monkeys and pigs are people punished by God for disobeying divine law'.[139] This belief is present in the Qur'an, which also states that some of the people turned into monkeys were Jews.[140] The Arabic terms for the baboons also had negative connotations. In formal Arabic, the baboons were called qurūd (singular qird), connoting 'a miser or one who lives on the cheap'. In popular speech, people called the hamadryas sa'dān, 'happy ones', or rubbāḥ, 'profiteers'.[141]

Despite their ambivalence, Arabia's hamadryas baboons attracted the attention of a few researchers. The German naturalists Christian Gottfried Ehrenberg and Wilhelm Hemprich, who explored the Red Sea region during the early nineteenth century, regularly saw baboons on Asir's mountaintops. After failing to shoot one of them in Arabia, they obtained an adult in the area of today's Eritrea. Ehrenberg also identified this species with the sacred baboon, the attendant of Thoth in ancient

Egypt.[142] During the 1960s and 1970s, Hans Kummer studied hamadryas in the Zurich Zoo and subsequently in Ethiopia. After the outbreak of Ethiopia's civil war in the 1970s, he gained the opportunity to continue his research in Saudi Arabia. It was his compatriot Wilhelm Büttiker, the instigator of the Zoological Survey of Saudi Arabia, who acted as an important link between Kummer and the Saudis. In addition, the Saudi zoologist Abdelelah Banaja facilitated Kummer's stay. The primatologist wrote, 'In 1980, thanks to Willi Büttiker's efforts as intermediary and to the generous help provided by Abdulelah Banaja of the University of Jeddah I was able to study the wild hamadryas baboons in Arabia for the first time.'[143] Banaja and other scientists organised drivers for Kummer's trips into the interior, accompanied him and served as translators.

Even during his brief fieldwork in Saudi Arabia, Kummer was able to notice the sensitivity of evolution. However, it did not seem to have been an important topic in his conversations with Banaja and other Saudi biologists. 'I was not in Saudi Arabia long enough to judge the extent to which science is accepted by religious teaching in Islam', Kummer wrote later. However, the primatologist noted 'tensions between science and religion in Islam'. 'Evolution is a difficult topic', stated Kummer, 'It is discussed, but also rejected.' Kummer also remarked upon the Islamisation of science that King Abdulaziz University, where Banaja worked, was attempting during the 1970s and 1980s: 'A Saudi colleague at the university, a biochemist, took a statement in the Koran as the subject of his research project, which he carried out with the most modern techniques.'[144]

Despite the sensitivity of evolution, Kummer gained important insights into the evolutionary history of the hamadryas through his field research in Saudi Arabia and Ethiopia. In 1995, he published a rare combination of an academic text and autobiography entitled *In Quest of the Sacred Baboon: A Scientist's Journey*. Published in English by Princeton University Press, it was neither primarily aimed at audiences nor underwent censorship in the Gulf. This makes it unsurprising that Kummer was able to devote an entire chapter to 'the likely evolution of hamadryas society'. In this chapter, he asserted the primacy of evolutionary biology for studies on animal behaviour. 'We see an animal species as the result of a gripping history during which characteristics emerged in the stream of external conditions', he stated. Kummer added that 'a species appears as an idea of survival – one way of coping with the world'.[145]

In contrast to creationists who emphasised divine action, Kummer asserted the importance of 'chance' in the generation of species. He wrote that the 'history of a species' adaptation can never be reconstructed with certainty, for it has been partly determined by the small events with great consequences that we generally refer to as "chance"'. Kummer stressed environmental factors in explaining the social structure of hamadryas baboons, including 'marriage'. He wrote that these baboons 'evolved in a region where food grew in such *small* and widely scattered areas that small foraging groups were necessary, and in these small groups marriage came into being'.[146]

Unlike many other biologists who worked in Arabia, Kummer did not limit himself to investigating the evolutionary origin of one particular species, but theorised about evolution in general. Engaging in a global debate rather than one restricted to the Gulf, he turned against an argument by the English sociobiologist and atheist Richard Dawkins. Dawkins had argued that plants, animals and humans were machines built by genes to propagate themselves. Kummer acknowledged that 'from the genes' point of view, that is right'. Yet, he stated that 'machines can do more than those who built them, or they would not have been built'. Similarly, plants, animals and humans have by far surpassed the genes that made them. This led Kummer to conclude that 'evolution is far more than the sorting of genes. When evolving species yield to the pressures of selection, they produce phenomena that leave the genes far behind in their kaleidoscopic richness and playfulness.'[147]

At the end of *In Quest of the Sacred Baboon*, Kummer also rejected that 'chance' is all there is to evolution. However, rather than turning to creation by God as a factor, he emphasised the importance of 'freedom'. This combination of evolution with a universal political concept further indicated that Kummer did not primarily aim at Gulf audiences, but a global readership. 'We are ungrateful when we dismiss the evolution of life as merely a cold game of chance with the macromolecules as dice', said the primatologist, 'It has been nothing less than a road to ever-higher levels of freedom.' Revealing a belief in a form of evolutionary progress, Kummer described freedom as 'emancipation from the rigid rules governing primitive forms of existence, with their single tracks, fixations, and blindness'.[148]

Although *In Quest of the Sacred Baboon* did not primarily target Gulf audiences, it was reviewed in Saudi Arabia. The NCWCD's magazine *Arabian Wildlife* reviewed Kummer's book positively, despite Kummer's theories about evolution, chance and freedom. The book 'has something for everyone – scientist, naturalist, armchair traveller, and student of human nature', wrote the reviewer. The review also referred to baboons' evolution. It quoted Kummer saying that the male baboon 'has evolved both of the fundamental aspects of fighting: a sharp canine tooth and a network of alliances'. Finally, the reviewer even referred to humans as a primate species and suggested proximity between baboons and humans. *Arabian Wildlife* stated that 'the fascinating thing about the patriarchal hamadryas males is that they have managed to integrate permanent female "relationships" into a cooperative male society, despite intense competition for females'. The review continued that this 'social structure is apparently found in only two other primate species', one of them being 'the human'.[149]

Although Kummer himself did not return to Arabia, research into baboons continued under the sponsorship of the National Commission for Wildlife Conservation and Development. This research did not focus on breeding hamadryas baboons, but, on the contrary, on controlling the increasing number of hamadryas that fed on the rubbish dumps on the outskirts of the cities. The baboons themselves thus became non-human actors in the networks of zoological research in the Gulf. In order to prevent an invasion of baboons into the cities, French and Saudi researchers at the National Wildlife Research Center, including Ahmed Boug, Jean-Pierre Gautier and Sylvain Biquand, embarked on a series of studies. During the late 1980s, they surveyed baboon populations from Medina to Ta'if, and discovered that troops of baboons persisted near urban centres and that these troops were larger than troops in the wild.[150] Subsequently, the scientists published several proposals on how to manage the growing numbers of baboons that had become dependent on food provisioning around cities. They included male vasectomy and hormonal control of female fertility, as well as the use of dogs and scare guns to repel baboons from crops and city outskirts.[151]

The French researchers did not theorise about life and evolution in a general way like Hans Kummer. However, they also wrote about primate evolution besides offering practical proposals on baboon control. Hence, not only primatologists visiting the kingdom, but also researchers employed in the kingdom were able to publish on the sensitive topic of evolution. In an article from 1994, Jean-Paul Gautier and Sylvain Biquand wrote that the impact of man was sometimes positive on primate demography. The primates' ability to switch to new food sources and to exploit man-modified environments were important for their 'evolutionary success'. The scientists also speculated that the adaption of primates to humans might have a long history. If contact between primates and humans was 'ancient', 'interactions between man and non-human primates are then not of a cultural nature but resort from the biology and co-evolution'.[152]

## Problems of subspecies

Evolution also entered the discussions at the National Commission for Wildlife Conservation and Development through taxonomy, which was particularly important for the conservation of gazelles. Because a variety of subspecies of gazelles existed, it was necessary to find the evolutionarily significant units, that is distinct populations for purposes of conservation. The quest for evolutionarily significant units was difficult, as researchers discovered, classified and reclassified new subspecies of gazelles. Often, single distinct specimens raised questions. A skull and skin from the Farasan Islands sent by Hemprich and Ehrenberg to Berlin in the 1820s differed from all other known gazelles. It received the name Arabian gazelle (*Gazella arabica*), but in the absence of other specimens, remained an 'enigmatic entity.'[153] In 1987, the governor of the province of Najran in southern Saudi Arabia donated another unique specimen, a small dark gazelle, to the NWRC. The biologists at the centre felt that this gazelle raised 'a taxonomic problem'. Reflecting disagreements among the scientists, the animal received different Latin names in the following years, including *Gazella gazella muscatensis, Gazella bilkis* and *Gazella gazella erlangeri*.[154]

The NCWCD considered the problems in finding the evolutionarily significant units important enough to organise an international workshop in Riyadh in 1992. The aim of the workshop was to produce guidelines and recommendations for research and conservation of gazelles in the kingdom. The NCWCD thus invited several scientists from foreign institutions as well as its own research centres. In their discussions, the participants repeatedly referred to evolution. The theory thus not only entered Saudi-sponsored scientific publications, but also academic debates in the Saudi capital. In one paper, Arnaud Greth and colleagues from the National Wildlife Research Center discussed the ongoing 'dispute' about the numbers of genuine species within the genus *Gazella*. Different scientists put the number between 12 and 16 each of which having subspecies. The NWRC researchers remarked that a subspecies 'may have a wide distribution with several scattered populations, each of which experiences different evolutionary pressures'. Greth and his colleagues thus asked, 'At which level should conservationists invest their efforts to conserve biodiversity?' In order to answer this question, the NWRC researchers considered research into taxonomy crucial. They warned that 'errors in systematics can lead to the recognition of taxonomic groups that share little evolutionary differentiation or, at the other extreme, to a lack of taxonomic recognition of phylogenetically distinct forms'. In both cases, efforts to conserve endangered taxa would fail to protect diversity.[155]

The paper by Arnaud Greth and his colleagues from the NWRC was not the only one referring to evolution. Douglas Williamson, a former director of the KKWRC, disagreed with the quest to find gazelle subspecies and preferred to preserve discrete populations instead. Yet, he also justified this preference in terms of evolution. Williamson suggested that 'every discrete population is on its own unique evolutionary track'. 'Once this track is terminated by extinction', he added, 'there is no way of putting a different group of animals onto the same track, whether or not they belong to the same race, subspecies or whatever.' Attempting to bypass the taxonomical controversy over subspecies, Williamson considered discrete populations as 'evolutionary units' and thus as 'units for conservation'.[156]

In the discussions of the NCWCD, evolution was thus not just relevant for the history of a species, but also for its future. Rather than focusing on past evolutionary processes and their results, that is subspecies, Douglas Williamson advised the NCWCD to focus on future evolution. He argued that human activities, such as hunting, disturbances of plant cover and the extermination of certain carnivores, created 'new selective pressures', which resulted in the 'emergence of new phenotypes' among Saudi antelope populations. Given the massive effects of human activities, Williamson considered it 'naive to believe either that it is possible to restore the past or that the future will be like the past'. According to him, one could only 'simulate what was previously present and to restore the ecological and evolutionary processes which generate diversity'.[157]

In addition to individual papers, evolution entered the concluding discussion about gazelle taxonomy at the NCWCD workshop. Arnaud Greth asked his colleagues why taxonomic questions should be raised in conservation and reintroduction programmes and which ones should be answered. Williamson responded that 'the goal of conservationists is to maintain biodiversity' and that 'taxonomy is one (and possibly the best) index of biodiversity'. Later in the discussion, Greth emphasised the importance of subspecies for conservation, arguing that they 'represent the evolutionary potential of new species'.[158]

The NCWCD's inquiry into taxonomy and evolution did not remain at the level of theoretical discussions. In order to investigate the taxonomy of gazelles further, the NCWCD established two genetic laboratories at the NWRC and the KKWRC between 1988 and 1991. These laboratories subsequently carried out a number of studies into the genetics of gazelles.[159] In 2001, researchers from the KKWRC and the Zoological Society of London led by Rob Hammond published a 'phylogenetic reanalysis' of the extinct Saudi gazelle, suggesting that it was an 'evolutionarily significant unit'. They relied on skins collected by St John Philby before 1940 and deposited at London's Natural History Museum. Using these specimens, the zoologists sequenced the genome of the Saudi gazelle in order to determine its position within the genus *Gazella.* They concluded that *Gazella saudiya* was 'distinct' from *Gazella dorcas*, and thus required conservation as a separate evolutionarily significant unit. *Gazella dorcas* was of African origin and 'unsuitable' for breeding and reintroducing *Gazella saudiya*.[160]

Although initially established for research into gazelles, the genetics laboratories of the NCWCD also investigated the evolutionary history of other species, most notably the hamadryas baboon. This genetic research solved problems surrounding the origins of the baboons. The German naturalists Ehrenberg and Hemprich and their successors, including Hans Kummer, had described baboons on both sides of the Red Sea, in today's Saudi Arabia and Eritrea. In his 1995 book, *In Quest of the Sacred Baboon*, Kummer found this geographic separation 'puzzling'. He considered it 'unlikely that two populations of savannah baboons would have evolved independently into identical hamadryas baboons, one on the horn of Africa and the other in Arabia'. Therefore, he concluded that 'the hamadryas must have originated on one side or the other'.[161]

In the 2000s, a small network of researchers in Saudi Arabia, France and Japan was able to solve Kummer's puzzle by analysing the genetic diversity of the baboons. Hammond undertook the initial genetic research in cooperation with the Saudi and French primatologists Ahmed Boug and Sylvain Biquand. In 2004, the scientists published an article in which they argued that 'baboons almost certainly evolved in Africa, rather than Arabia, given the much greater diversity and range of baboon species in Africa compared to Arabia'.[162] A couple of years later, Takayoshi Shotake, a Japanese primatologist from Kyoto University continued this research. Working closely with Ahmed Boug, Shotake produced a new hypothesis. He suggested that during a period of glaciation, ancestors of modern baboons crossed the straight from East Africa to Arabia. 'The hamadryas then evolved its distinctive characteristics in Arabian isolation over hundreds of thousands of years', the Japanese primatologist suggested, before it 'crossed back into the East African mountains'.[163]

Although mainly involving research centres in Saudi Arabia, genetic research on Arabian mammals also received support from institutions in the smaller Gulf states. One of them was Al Wabra Wildlife Preservation in Qatar, an institution that also started as a farm belonging to a member of the ruling family, similarly to the King Khalid Wildlife Research Center. Sheikh Saud bin Muhammad Al Thani, who inherited the farm from his father, turned it into a research and breeding centre for endangered species, including antelopes and birds. His collection provided living gazelles, complementing the museum specimens in Europe. The 'phylogenetic reanalysis' of the Saudi gazelle led by Rob Hammond relied on samples from Sheikh Saud's antelope collection and Al Ain Zoo in the United Arab Emirates in addition to Philby's skins from the NHM.[164] In another study, researchers based in Cambridge, Munich and Al Wabra reinvestigated the enigmatic Arabian gazelle. Performing a phylogenetic analysis of the skull and skin collected by Ehrenberg and Hemprich and kept at Berlin's Museum für Naturkunde, they found the skull and skin belonged to different animals. Each of them were part of distinct lineages of the mountain gazelle. Rather than forming a separate and extinct species, Arabian gazelles thus formed an extant clade of mountain gazelles.[165]

## Conclusions

From the 1950s onward, the Gulf monarchies invested significantly in the conservation of wildlife, establishing zoos, wildlife reserves and research centres. These efforts initially focused on certain flagship animals, such as the Arabian oryx, in which the princes had particular interest. However, within the patronage networks of rentier science, researchers retained considerable agency. They were able to broaden the conservation programmes to include other species of mammals, such as gazelles and hamadryas baboons. They were also able to gain governmental support for research into animals with less symbolic value, such as insects. Finally, yet importantly, the scientists retained sufficient freedom to use the sensitive theory of evolution when explaining Arabian biodiversity, orxy survival, primate behaviour and gazelle taxonomy.

Patronage networks between the princes and scientists, as well as collegial networks between researchers based in Arabia and Europe were of pivotal importance to much of the conservation research in the Gulf. In return for expertise and environmental credentials, the princes provided scientists with considerable funding and infrastructure for the reintroduction of animals and associated research. In exchange for funding, access to living animals and data, colleagues in Europe provided expertise and museum specimens. Western institutions also provided researchers in Arabia with opportunities to publish beyond the reach of the Gulf's censorship agencies.

Zoological research, however, did not only encompass insects and mammals. Of at least equal importance to the Gulf rulers as antelopes were birds. The next chapter will explore the research that ornithologists pursued in the Gulf, and the networks that enabled them to do so. The chapter will also explore why the creationist princes or Gulf governments did not effectively censor evolutionary theory in the publications produced under their patronage.

# CHAPTER 4

# SCIENTIFIC ISLANDS OF EFFICIENCY

While efforts to conserve the Arabian oryx and gazelles gained much support from the princes, the houbara bustard, the most important prey in Arabian falconry, received even more attention. During the twentieth century, houbara numbers decreased dramatically because of over-hunting. Hunters used not only falcons, but also shotguns and other modern weapons to which the houbaras were not adapted. The declining numbers of houbaras threatened falconry as a sport and part of the monarchies' national heritage. Research on houbaras was thus at the centre of conservation efforts in the Gulf. The main goal behind the establishment of the National Wildlife Research Center in Saudi Arabia during the mid-1980s was not the conservation of the Arabian oryx, but the breeding and reintroduction of houbaras.

Senior members of the Gulf's ruling families were at the centre of the financial networks that sustained research on houbaras. Besides the House of Saud, which funded the NWRC, the House of Nahyan in Abu Dhabi supported houbara research for decades. In 1976, Zayed bin Sultan, the ruler of Abu Dhabi, sponsored the first International Conference on Falconry and Conservation in his emirate.[1] It resulted in the establishment of a houbara breeding programme in the late 1970s.[2] In 1993, a cabinet resolution banned the importation of houbara bustards, a measure in line with the Convention on International Trade in Endangered Species.[3]

As part of their conservation efforts, members of the House of Nahyan, like their peers elsewhere in the Gulf, established strong connections with international organisations. This helped local rulers win environmental credentials, while securing the future of hunting. Conservation initiatives at first helped against criticism of unsustainable hunting practices across Asia and Africa by sheikhs from the Gulf. Later, environmental credentials also became important as the Gulf monarchies came under increasing criticism for their carbon dioxide emissions, which were some of the highest in the world.[4] Ironically, with the money available through the hydrocarbon industries, the Gulf rulers were able to associate themselves with, if not co-opt, international environmental organisations. Abu Dhabi housed a Houbara Specialist Group that belonged to the International Union for the Conservation of Nature.[5] In 1995, Zayed bin Sultan received an award from the United Nations Food and Agriculture Organization 'in recognition of his achievements in greening the country's deserts'.[6] In 1997, the WWF awarded Sheikh Zayed the Gold Panda, the highest award in recognition of donors.[7] In 2001, his son Hamdan patronised the establishment of the Emirates Wildlife Society in association with the WWF.[8] In 2005, the United Nations Environment Programme posthumously awarded Zayed bin Sultan its Champions of the Earth award.[9]

The Gulf governments used international recognition for domestic legitimation too, as 'sustainability' became an important part of their political rhetoric. Environmentalism thus also fitted with the policies to diversify their economies.[10] *The National* commemorated Sheikh Zayed's winning of the Gold Panda in 1997 as part of its coverage of the United Arab Emirates' fortieth anniversary. The newspaper quoted a member of the executive committee of the Environment Agency – Abu Dhabi saying, 'For five decades, Sheikh Zayed advocated and adopted the concept of what is known today as sustainable development.' The official added, 'Various developments across the UAE have shown that while there was a great interest in improving the social and economic conditions of the people, great care and importance was given to protect the environment.'[11]

Experienced in dealing with, if not co-opting, international environmental organisations, Abu Dhabi even managed to gain the support of the United Nations for hunting practices. In 2010, the United Arab Emirates persuaded the United Nations Educational,

Scientific and Cultural Organization to list falconry as an Intangible Cultural Heritage to Humanity.[12] In its justification, UNESCO suggested that falconry was environmentally friendly. The committee concerned decided that falconry 'is a social tradition respecting nature and the environment, passed on from generation to generation, and providing them a sense of belonging, continuity and identity'.[13]

Falconry and houbara conservation was not only a personal interest of individual rulers, like Zayed bin Sultan, but also a family affair. After Sheikh Zayed's death in 2004, one of his sons institutionalised patronage for houbara research. In 2006, Mohammed bin Zayed, the Crown Prince of Abu Dhabi, established the International Fund for Houbara Conservation (IFHC) in order to secure the future of falconry. Mohamed Al-Baidani, the fund's director general, warned that 'as long as the Houbara is at risk, the future of traditional falconry is also at risk.' He thus described the fund's programmes as 'the best hope for the preservation of the Houbara in the wild'. 'In turn,' Al-Baidani added, 'preservation of the Houbara helps maintain our traditions and brings the balance back to our environment realising Sheikh Zayed's vision.' By 2011, IFHC-funded programmes had produced around 90,000 houbaras, and aimed at producing, breeding and releasing 50,000 birds annually.[14]

In order to produce houbaras quickly, senior members of the House of Saud and the House of Nahyan established the NWRC in Saudi Arabia and the National Avian Research Center (NARC) in Abu Dhabi in the late 1980s. Under royal patronage, these centres formed scientific 'islands of efficiency' within the larger rentier bureaucracies. They were similar to certain dynamic, profitable and growing state-owned enterprises, such as the Saudi Basic Industries Corporation (SABIC) or Emirates. In contrast to other parts of the public sector, the centres received patronage and orders from a limited number of senior princes and were largely insulated from wider rent seeking. Their managers were hired based on merit and had substantial autonomy.[15]

Although the scientific islands of efficiency received clear mandates – to produce houbaras – their autonomy and protection also allowed them to engage in wider areas of research. Foreign managers recruited scientists mainly through international channels. Offering competitive salaries and funding for research, they attracted ambitious and internationally mobile scientists who were interested in broader aspects of desert ecology. These scientists used the research facilities and environments of the Gulf to move into areas of research unintended by their patrons, including evolution and climate change. However, in contrast to companies, such as SABIC or Emirates, the scientific islands of efficiency did not have a source of income independent from their patrons. This made wildlife research centres even more reliant on individual princes.

## From breeding to desert ecology research

The first major hub for houbara conservation and research in the Gulf monarchies was the NWRC near Ta'if. It was similar to some of the Gulf's highly efficient state-owned enterprises through its employment of highly paid foreign experts.[16] Funded and supervised by the National Commission for Wildlife Conservation and Development, the NWRC was run by Jacques Renaud's company Reneco. In order to develop a breeding programme, Renaud used international connections to hire most of the centre's staff. In 1989, Renaud recruited a German bustard specialist named Holger Schulz as director of the NWRC. Previously, Schulz had completed a doctoral dissertation on the little bustard and had become co-chair of a working group on bustards that was part of the International Union for Conservation of Nature.[17]

Initially, the main aim of the NWRC was the breeding of houbara bustards, but Holger Schulz was able to broaden the scope of research. When he first arrived, he found the Saudis mainly interested in producing 'as many houbara bustards as possible' for reintroduction into the wild. His mandate was thus to develop artificial insemination, to produce 'as much offspring as possible' and to raise the chicks. However, as he investigated possible locations for releasing the birds, he found that many of them had been 'destroyed and overgrazed by too many herds'. The ornithologist thus instructed his staff to study houbara behaviour in their habitat and to 'find ways to make areas available in which the houbaras could live'. This required not just studies on artificial insemination, but also 'field research' and 'physiological research'.[18]

Benefitting from the interest of Saud Al-Faisal and other princes in houbaras, Schulz quickly expanded the NWRC's research into houbara ecology. Important for his success was the financial backing of

Saud Al-Faisal and the NCWCD, which funded the centre through Reneco. Compared to 'normal research' in Germany, Schulz found the funding situation in Saudi Arabia to be 'paradise-like'. He still had to 'fight for money' in order to hire five new scientists for the centre's expansion. Yet, despite some budgetary constraints, Schulz found that if he convinced his Saudi patrons of an initiative, 'the funding always became available.'[19]

Similar to highly efficient state-owned companies, personal connections to members of the ruling families were crucial for receiving support. However, the NWRC was not able to bypass the national bureaucracy completely. Instead, Schulz had to work through the NCWCD, which began as a small support agency for the centre, but over time turned into a larger, more bureaucratic and clientelistic, entity. Schulz said that there were 'always small frictions' between the consultant manager Jacques Renaud and Abdulaziz Abuzinada, the secretary general of the NCWCD. Schulz needed the NCWCD, because 'a French team could not have worked in Saudi Arabia without the Saudis being fully behind it'. In the end, Schulz was able to gain 'much support' from Abuzinada and Saud Al-Faisal, the managing director of the NCWCD. An important intermediary linking the foreign scientists and the leadership of the NCWCD was the Saudi botanist and client of Abuzinada Abdul-Rahman Khoja. Schulz described Khoja as 'the contact to Dr Abuzinada, to Prince Saud', adding that 'without good personal connections, the whole thing would not have worked as well as it did'.[20]

Despite the Saudi financial support for his institution, Schulz's family did not adapt to the Saudi environment. In his case, gender norms challenged the international recruitment of a scientist in the Gulf. Schulz left the National Wildlife Research Center in 1991, two years after his appointment as director. On the one hand, Saudi funding had enabled him to lead ornithological research that was more ambitious and expensive than in Germany. On the other hand, his family had known from the start that they 'would not stay forever' in the kingdom. Schulz also felt that his family was particularly isolated at the camp of the NWRC. This camp was outside the city of Ta'if, and his wife was not allowed to drive a car or to be 'away on her own'. While his wife and two children 'effectively could not get out' of the camp, Schulz himself often left for trips to various parts of the kingdom. Family grievances thus encouraged Schultz to accept the directorship of a research institute back in Germany.[21]

Before Schulz left, he used international networks to recruit a few more scientists to work on desert ecology. In 1990, Schulz met Georg Schwede at an ornithological conference in Germany and hired him as manager of the recently established Mahazat as-Sayd Protected Area near Ta'if. Upon Schulz's departure from Saudi Arabia, Jacques Renaud, the consultant manager, made Schwede the NWRC's new director. After two years, however, the networks between the Gulf and international organisations also provided Schwede with an opportunity to leave Arabia. From the NWRC, he moved to WWF Germany and soon became its managing director.[22] Subsequently, he became a director of WWF International in Switzerland.[23]

Despite the high turnover rates of foreign staff, strong international networks and the high level of funding allowed for quick replacement of departing staff. Besides Schwede, Schulz hired a number of other internationally mobile scientists.[24] They included an ornithologist from New Zealand named Phil Seddon and his partner Yolanda van Heezik, who were doing postdoctoral research on South African penguins. Seddon and van Heezik worked at the Percy FitzPatrick Institute of African Ornithology, one of the main hubs of ornithological research in Africa. However, as their contracts were temporary, Seddon was 'applying for any and all suitable jobs'. They responded to an advertisement in the magazine *New Scientist* recruiting biologists to survey birds in Saudi Arabia. Van Heezik received a letter informing her that single women could not work in this position. Schulz invited Seddon, however, for an interview in Kenya as a third country, as Saudi Arabia only established diplomatic relations with post-apartheid South Africa in 1994. Following the interview, Schulz asked Seddon to 'come immediately'. After a 'quick wedding' in Cape Town, Seddon and van Heezik landed in Jeddah in November 1991.[25]

Despite the arrival of new scientists, the departure of Holger Schulz as director had a negative impact on the centre's research. As the patronage for the centre relied on personal connections, the disruption of this connection also threatened funding. During his tenure, Schulz had persuaded Saud Al-Faisal to expand the budget of the NWRC considerably. Schulz had envisioned the centre to become a major desert research institution similar to centres in Israel or India. However, after

Schulz left, the newly arrived researchers were unable to 'fight' for the budget. 'All the big plans,' said Seddon, 'collapsed.' This was during a time when oil prices and thus Saudi revenues had declined significantly from a peak around 1981. The Gulf War against Iraq between 1990 and 1991 further burdened Saudi finances. Alongside other government agencies, the NCWCD was thus cutting budgets during the 1990s.[26]

In this financial situation, the NWRC focused its work on the core objective of houbara breeding, which was of most interest to the princes funding the centre. Seddon said that the centre's 'success' at the time was 'measured in terms of captive-breeding output of houbara bustards'. As an entrepreneur, Jacques Renaud was mostly concerned with increasing the production of captive houbaras year after year. Any research that was not seen as directly contributing to this 'was of lower priority'. Despite the high profile of the Arabian oryx programme, Seddon speculated that 'the centre would have collapsed' without the production of sufficient houbaras. 'The funding would have dried up.'[27]

Thanks to scientific expertise and animals drawn together from different continents, the NWRC succeeded in breeding houbaras. As houbaras had almost disappeared from Arabia, the NWRC sent expeditions to Pakistan and Algeria between 1986 and 1988. With permission from local authorities, the expeditions collected fertile eggs of the Asian and African subspecies of the houbara, which formed the basis for an initial breeding stock of more than 200 chicks.[28] Subsequently, the NWRC produced houbaras by means of artificial insemination and started to release them in reserves. Between 1991 and 2010, more than 800 houbaras entered the Mahazat as-Sayd Protected Area alone.[29]

Although the NWRC did not expand its programme according to Schulz's ambitious plans, the employment of researchers led to the production of scientific papers alongside the breeding activities. Jacques Renaud 'didn't really care' about scientific publications, according to Seddon. In order to secure the renewal of the contract with his company Reneco, Renaud was more interested in reports and other outputs that would satisfy Saud Al-Faisal. These reports included observations by Seddon and van Heezik of a population of houbara bustards in the Harrat al-Harrah Protected Area in northern Saudi Arabia. 'As long as we were up there, doing counts, producing reports and showing that we were trying to understand more about the small population up there, everyone was happy', said Seddon. He added that although 'there was no pressure to publish,' he and his wife 'felt as scientists it was important to keep publishing.'[30] Over the years, Seddon and van Heezik were able to broaden their research beyond houbaras to include other desert birds, wolves, small mammals and vegetation. Although some of this research was of little 'direct relevance' to houbaras, Seddon found that it still enjoyed 'some degree of support from Riyadh'.[31] This support was probably related to the broader vision of the botanist Abdulaziz Abuzinada to conserve Saudi Arabia's entire biodiversity instead of a single species.

Seddon and van Heezik did not focus on evolution, but they did not avoid it either. Thus, while concentrating on breeding, the NCWCD inadvertently funded studies in the evolutionary history and adaptations of desert animals. A paper co-authored by Seddon discussed the fossil record of Rueppell's sand fox, one of the predators threatening houbaras. Seddon wrote that 'desert adaptations evolved at least twice' in the larger biological family of foxes and dogs. Rueppell's sand fox 'diverged' from other foxes relatively late, 'suggesting that it entered desert regions more recently'.[32] In another article, van Heezik and Seddon described wild houbara bustards as 'superbly adapted to arid environments'. The birds did not need to drink water but gained all the necessary moisture from their food. The scientists suggested that a varied diet including berries, beetles and lizards formed the 'secret of the bustards' survival'.[33]

In order to carry out their research, Seddon and van Heezik relied not only on the support of government agencies, such as the NCWCD, but also on the Bedouin. As in the case of the research by the Egyptian botanist Kamal Batanouny, nomads thus formed an important part of the networks of Gulf biologists. Without them, Seddon and van Heezik would have felt 'isolated' in the desert area of Harrat al-Harrah.[34] Unfamiliar with local landscapes, they described the place as rather extra-terrestrial, similar to British colonial environmental imaginaries of Arabia.[35] The Harrat al-Harrah Protected Area covered an ancient lava field that looked 'like the surface of Mars' to Seddon. The researcher from New Zealand said that he 'loved the desert' with its 'stillness' and 'unique flora and fauna'. Yet, he also feared it 'because the usual safety networks and links were absent'. In this situation, Seddon and van

Heezik, who was learning Arabic, interacted with the local Bedouin rangers 'a great deal'.[36]

A sense of extraterritoriality added to the extra-terrestrial appearance of northern Saudi Arabia. As the research centres formed islands of efficiency within the larger rentier bureaucracy, the reserves formed islands within the Saudi legal and social system. Harrat al-Harrah protected not only animals against human interference, but also humans against the gender regulations of the Saudi state. Seddon and van Heezik often stayed at the Bedouin rangers' camp and found them to be 'very hospitable' and 'accepting'. Initially, van Heezik wore the abaya, the long overgarment that women were required to wear in public in Saudi Arabia. After a year, however, the rangers told her 'not to bother', as they would think of her 'like a sister'. The Bedouin not only provided company, but also knowledge, despite having limited literacy and formal academic training. Seddon emphasised that his 'ability to conduct research on the desert fauna benefitted hugely from the deep knowledge and extraordinary tracking abilities of the rangers'.[37] As was the case with botany, zoology in the Gulf thus combined 'local' or 'indigenous' and 'expert' knowledge.[38]

In order to publish articles in the relative isolation of Harrat al-Harrah and the NWRC, Seddon and van Heezik relied on their connections with colleagues abroad. Global as well as local networks were thus important for zoological research even at remote locations, such as the northern Saudi deserts. The two ornithologists benefitted from a 'small library' at the NWRC. More important, though, were friends and colleagues abroad who often sent them papers. When Seddon and van Heezik were outside the kingdom, they also visited university libraries in order to 'catch up on literature'. As the 1990s progressed, the staff of the NWRC recognised that they 'needed the internet'. Van Heezik came to realise that 'everyone else assumes that you have the internet' and 'all of the kind of professional networking was done' through it. Before the Saudi government introduced public access to the internet in 1999, the centre thus used a radiotelephone to connect to a server in Bahrain.[39]

Ultimately, many of the foreign researchers failed to adapt to the Saudi environment, unlike the animals they were studying. In 2000, Phil Seddon and Yolanda van Heezik left Saudi Arabia, as the Germans Schulz and Schwede had done before them. As had been the case with

Schulz, the couple's reasons related to their domestic situation and Saudi gender norms and laws. By the time Seddon and van Heezik left, they had a four-year-old son. They were unhappy with sending him to an international school in Ta'if and 'didn't want him to go to a Saudi school' either. Saudi laws prevented van Heezik as a woman to drive him to school, and she was not willing to entrust her son to a foreign driver.[40] Besides family reasons, Seddon was concerned about his precarious situation as a foreigner in the Gulf – even as a highly paid expert. Like many other expatriates in the Arab oil states, his material life was comfortable, but his emotional life was 'wanting in security and lacking a sense of belonging'.[41] In the nine years that he was in the kingdom, 'the sense of being a foreigner never went away', he said. 'I was very much aware I was there to do a job and that once that job was done, or when a Saudi could be found to do the job, then I was to leave.'[42]

The departure of foreign scientists like Seddon and van Heezik left a gap in the houbara research, as only few Saudi nationals were able to replace them. In contrast to successful state-owned enterprises, such as SABIC,[43] the NCWCD's recruitment of Saudi citizens relied more on patronage rather than merit. Consequently, the NWRC did not promote enough Saudi employees to a level where they could successfully take over the organisation. Often, Saudis did not join the centre through competition, but the NCWD had 'prescribed' their employment. In Schwede's view, many Saudis 'did not have the capacity and maturity to hold certain positions of responsibility'. At the same time, the expatriate scientists were 'under pressure' to produce successes in breeding houbara bustards. The centre's leadership thus excluded Saudi novices from 'strategic decisions' and the 'management of larger programmes'. Looking back, Schwede stated that he should have put local 'capacity building' as one of the main tasks in the job descriptions of the expatriate scientists. In its absence, Schwede considered 'only external know-how, which had been purchased globally' as responsible for the quick 'successes' of the centre.[44] In the case of the NWRC, the international scientific networks in combination with the effects of a patronage-based bureaucracy thus resulted in a failure to develop scientific expertise among Gulf citizens.

Although the heavy reliance on foreigners in combination with patronage-based employment of nationals hindered local capacity building, it was a wider pattern in science in the Gulf monarchies. Abu Dhabi followed Saudi Arabia's model of producing houbaras quickly, using some of the same internationally mobile scientists. In 1989, Zayed bin Sultan, the ruler of Abu Dhabi, decreed the establishment of a National Avian Research Center, bringing professional ornithologists for permanent work to the UAE for the first time.[45] Some of these ornithologists had previously worked at the NWRC in Ta'if and were thus able to transfer their expertise to the NARC. One of them was a Frenchman named Fred Launay, who had undertaken research on houbaras at the NWRC together with Patrick Paillat and Arnaud Greth.[46] In the mid-1990s, Launay moved to NARC and soon became its director.

As in the case of the NWRC, the main activities of the scientists at the NARC included breeding, releasing and monitoring houbaras. These activities ultimately aimed at resurrecting local houbara populations for hunters and diverting them from hunting free-ranging populations in Central and South Asia.[47] NARC's activities did not involve research into evolution per se. However, scientists like Launay referred to the theory in their papers. In a report entitled *Counting Houbara Bustards*, Launay reflected on the difficulty of finding birds that were hiding from falcons and other predators in the Arabian wilderness. 'The task is complicated for any living species', Launay stated, but it was even more difficult for 'a species which for survival and through evolution has come to depend mainly on its ability to avoid detection'.[48]

Occasional references aside, several scientists at the NARC also worked more closely on evolution. Princely patronage thus enabled them to undertake research on a sensitive topic alongside less controversial conservation activities. One of the scientists at NARC interested in evolution was the British ornithologist Simon Aspinall. A graduate in environmental science from the University of East Anglia, Aspinall had worked for nine years for the Royal Society for the Protection of Birds, before joining NARC in 1993.[49] In his research, he teamed up with Phil Hockey, a British researcher at the Percy FitzPatrick Institute of African Ornithology in South Africa.

Together with Phil Hockey, Simon Aspinall researched the evolution of the crab plover, a bird that lived on the shores of the Indian Ocean in Arabia and East Africa. In an article from 1996, Hockey and Aspinall described the singular evolutionary status of crab plovers. They wrote that the lineages of the crab plovers and their closest relatives 'diverged during the Oligocene, leaving the Crab Plover on an evolutionary monorail for the past 35 million years'. The ornithologists also explained why crab plovers excavated their own underground nest burrows. They stated that 'underground breeding has evolved in many animal groups, including birds, as a means of escaping from predators'. However, in the case of the crab plover, the researchers suggested that 'the reason for the evolution of burrow-nesting is more likely avoidance of heat than of predators'.[50] In 1997, the two scientists published another article explaining why the crab plover only laid a single, large egg. They suggested that this egg was 'an adaptation to minimising brood energy demand' and 'to reducing fledging period'. Minimising the amount of energy spent on breeding was particularly important when crab plovers were 'energetically stressed'. In such a situation, the ornithologists considered the laying of a single egg as an 'evolutionarily adaptive solution'.[51]

Despite the sensitive nature of evolution, there is no evidence that these evolutionary references caused any problems for Aspinall's work in the Gulf.[52] Together with Phil Hockey, he was even able to write about his research in the Saudi popular magazine *Arabian Wildlife*. In an article from 1997, Aspinall and Hockey repeated their claims about the crab plovers' 'isolationist evolutionary history'. 'The crab plover has been on an evolutionary monorail for some 35 million years', they stated. The ornithologists stated that 'the evolutionary eccentricities of the unique crab plover are only just beginning to come to the surface'. According to the researchers, much more work was necessary so that 'their survival for the next 35 million years can be assured'. The researchers explicitly warned that 'oil pollution is an ever present problem'. They thus problematised a sensitive industry as well as using a sensitive theory in the Gulf.[53]

Foreign experts, like Aspinall, continued to lead most of the work at the NARC during the 1990s and 2000s. As in the case of the NWRC, scientific capacity building among Gulf nationals remained limited. Despite the popularity of falconry in the UAE, only a few Emiratis joined NARC. One of them was a former post-office clerk who moved to the centre in 2008. Driving a 4x4 vehicle through the dunes of a protected area, his new job was to monitor how the released houbaras were surviving. However, his case was still rare enough to be worthy of a lengthy article in the newspaper *The National* in 2013. This article itself probably sought to attract more Emiratis to science and conservation. Celebrating an achievement under princely patronage, the article ended by stating that the House of Nahyan's International Fund for Houbara Conservation had 'released 1,400 birds in the UAE during the last season'.[54]

## A turn to physiology

Like the NARC in Abu Dhabi, the NWRC in Saudi Arabia mainly employed foreign researchers during the 1990s and early 2000s. While many of these researchers were transitory, a number of them stayed long enough to become important nodes linking researchers based in the kingdom and abroad. One of them was the French veterinarian Stéphane Ostrowski, who arrived in the kingdom through a network that included not only scientists and business, but also the postcolonial French state. As the British Empire withdrew from the Gulf during the decades after Indian independence, the French state developed stronger political and economic interests in the region and supported cultural and scientific exchange with the Gulf states. This support included the partial funding and administration of positions in scientific institutions in the Gulf.

Ostrowski gained this position under the supervision of the French embassy in Saudi Arabia at the NWRC in 1993, after graduating from the École nationale vétérinaire de Lyon. This veterinary school, the oldest in France, had long produced veterinarians for the French Empire in North Africa and elsewhere.[55] As a 'civil collaborator', Ostrowski undertook voluntary service in Saudi Arabia instead of military service in France. Convinced by an idea of peaceful international development cooperation, he preferred to help 'a foreign society at learning something useful' instead of serving in uniform. For 16 months, Ostrowski formally served under the French cultural attaché in Riyadh, but in practice under the French head veterinarian at the NWRC, who was at first Arnaud Greth and later Marc Ancrenaz, two other graduates of French veterinary schools.[56]

The high turnover rate of foreign experts and the lack of training of Gulf nationals offered opportunities for young Western professionals, such as Ostrowski, to realise their ambitions in the Gulf. In 1993, Arnaud Greth left the NWRC after four years to join – like Georg Schwede – the WWF. Marc Ancrenaz quit his position as head of the NWRC's veterinary department after less than three years in 1995 for an orang-utan conservation programme in Malaysia. At the end of his 16 months as civil collaborator, Stéphane Ostrowski decided to take his two months of end-of-duty vacation in Saudi Arabia. Ostrowski claimed that he 'developed a true love for deserts and the desire to understand how life was possible in this extreme environment'. He soon received an offer to succeed Ancrenaz as head of the NWRC's mammal and veterinary departments. Ostrowski accepted this offer on the condition that he could expand his duties beyond veterinary science and houbara breeding. He was particularly interested in developing research on the physiology of Arabian oryx and gazelles.[57]

Ostrowski subsequently became a central figure in the management of research on desert ecology at the NWRC. He was thus important for broadening the centre's activities from breeding into areas more closely related to evolution. In 1995, he started as head of the mammal and veterinary departments at the centre. During the same year, he also married and was joined by his wife. A trained veterinarian herself, she worked as administrative officer at the NWRC and later as manager of the archive and library resources. A year later, Ostrowski had acquired sufficient knowledge and management capacity in order to venture beyond daily activities into the long-term planning of research.[58]

One of the results of this planning was a turn of the NWRC in the direction of physiology. This took place in tandem with developments at the NWRC's sister centre, the King Khalid Wildlife Research Center. In the mid-1990s, Jacques Flamand, one of Ostrowski's colleagues, became director of the KKWRC with the aim of developing genetic research. In this case, the vertical networks between the French staff members at the scientific islands of efficiency allowed for more communication than was common between other Saudi governmental agencies (such as MEPA and the MAW). In order not to duplicate efforts and create 'unnecessary competition', Ostrowski avoided a research initiative in genetics. Instead, the veterinarian considered research on the eco-physiology of desert ungulates 'both useful for conservation and scientifically meaningful'. Because he had limited knowledge in this field, he sought to attract new collaborators.[59]

Fortunately, the ornithologists Phil Seddon and Yolanda van Heezik at the NWRC had connections with physiologists. From their time at the Percy FitzPatrick Institute of African Ornithology, they knew the American ornithologist Joe Williams, who had worked in South Africa during the 1980s and early 1990s. During that time, Williams had already studied the energetics of birds with reference to evolution. In an article written with a colleague, he had found that green wood hoopoes from South Africa conserved energy by roosting in groups in cavities. In areas where temperatures sometimes drop below freezing during winter nights, this behaviour saves lives. Williams and his colleague suggested that the conservation of energy 'may have been important in the evolution and/or maintenance of sociality in this species'. In another study, Williams found that birds adapted to arid environments had a reduced water loss even at lower temperatures. Williams concluded that 'selection has operated to reduce water loss in these species' even at moderate temperatures.[60]

Activating her ornithological network from South Africa, Yolanda van Heezik recommended Williams as a partner in research on Arabian oryx energetics. Afterwards Ostrowski contacted Williams, who had since joined Ohio State University, via fax and email, and the two scientists soon established trust. Ostrowski 'immediately felt in very close agreement with Joe's research ideas and philosophy of life'. The veterinarian added, 'I just knew that he was the one who would help me develop the research activities I envisioned.'[61] What helped build the trust was the quick support from the NWRC as an island of efficiency within the Gulf's large public sector. Without having met the French veterinarian, Williams told Ostrowski that a project on the water management of Arabian oryx required isotopes worth $10,000. 'And the next week I had a check for $10,000', Williams said. The American zoologist immediately bought the isotopes and flew to Saudi Arabia.[62]

Adding to the NWRC's connections, Joe Williams brought his own scientific network to Arabia. This contributed to making the NWRC a hub of transnational research similar to the Percy FitzPatrick Institute of African Ornithology. Williams involved his own graduate students, including Irene Tieleman, in his research in Saudi Arabia. A student at the University of Groningen, Tieleman had initially planned to work on birds in southern Africa. However, after Williams's first visit to Saudi Arabia in 1996, he invited Tieleman to accompany him. Under

Williams's supervision, Tieleman undertook two physiological projects for her master's degree and doctorate. Her master's thesis, completed in 1997, built on Williams's research and investigated the role of nasal membranes in reducing evaporative water loss in desert larks. Tieleman's subsequent doctoral project was entitled 'the ecophysiology of species of larks along an aridity gradient'.[63] She found the family of larks particularly suitable for such a study, as it had members in environments ranging from arid to semi-arid, mesic and even arctic areas.[64] In Saudi Arabia, Tieleman was able to study larks adapted to a particularly high degree of aridity.

The study of physiological adaptations of birds required Tieleman to make her own adjustments to the social environment of Saudi Arabia. These adjustments were not as major as for women employed full-time in the kingdom, like van Heezik. Nevertheless, state-enforced gender segregation made the integration into male-dominated scientific networks difficult for Tieleman. As a feminist, she initially objected to wearing an abaya, but Williams persuaded her to do so at least outside the reserves.[65] The Dutch researcher nevertheless remembered 'attracting attention from everyone anyway', as she did not cover her head and was taller than most local men. On a ferry, she had to sit in what she called the 'claustrophobic' women's section, after a member of the Committee for the Promotion of Virtue and the Prevention of Vice had denounced her to the captain. Several local conferences also excluded her, while inviting her male colleagues.[66]

In the reserves of the NCWCD, Tieleman had more freedom. The protected areas thus formed a space where humans were also set apart from wider society and its conventions. Like the British botanist Sheila Collenette – and nineteenth-century women explorers[67] – Tieleman was considered not completely female during fieldwork. She was not allowed to drive cars on public roads in Saudi Arabia, but did so 'with impunity' in the Mahazat as-Sayd Protected Area.[68] For the local rangers in the reserves, Williams said, 'she was Dr Tieleman', a 'scientist', and 'she wasn't really considered to be a female'. She even ate *kabsah*, a common rice dish, with the Bedouin rangers. 'Anything I was invited to by the rangers, she was invited to, for tea, for coffee, food, anything', Williams said, 'She was considered an equivalent.' From the rangers, both Williams and Tieleman slowly learnt the local vernacular. 'We learnt Arabic essentially phonetically, without knowing the alphabet very well', Williams said. 'We were just writing down words from the way they sounded and used them.'[69] When it came to Arabic, the Bedouin thus possessed 'expert knowledge', while the foreign scientists were initially almost illiterate.

**Figure 4.1**   Joe Williams in front of a truck of the NCWCD in 2007 (courtesy of Joseph B. Williams)

Despite the restrictions on women in Saudi law, the NWRC supported both Tieleman and Williams. Quick money from a scientific island of efficiency in the Gulf also allowed the Western researchers to start projects that were difficult to realise in the more competitive and bureaucratic funding environments in the West. In 1998, the NWRC provided Williams and Tieleman with $11,000 for supplies and equipment to study the 'evolutionary physiology of larks in Saudi Arabia'. This was followed by $54,000 for supplies and equipment for a project on 'Ecophysiology of Oryx in Saudi Arabia' between 1999 and 2003.[70] This money was especially important, as Williams initially had limited funding from America's National Science Foundation. 'The NSF wouldn't fund us', he said, 'because they did not think the problem was important enough or that we could do it.' Only 'three or four years into the project', the NSF realised that his team was 'really doing some good stuff'. In 2002, the NSF awarded him a grant of $254,000 for a project on 'skin lipids and cutaneous water loss among larks along an aridity gradient'. This grant paid for their airfare to Saudi Arabia and equipment costs, while the NWRC paid for expenses inside the kingdom.[71]

Apart from funding, the NWRC provided the foreign researchers with much practical and logistical help, including trucks, fuel and food.[72] Ameliorating the effects of gender segregation, Stéphane Ostrowski acted as an important intermediary between Tieleman and local male staff. This mediation made the centre a kind of third space in terms of social norms, a transnational space in between the Gulf and the West. Tieleman found that many locally hired staff did not take her 'very seriously'. Drivers who brought supplies from Ta'if to the centre accepted orders from Ostrowski, but not from her. However, if she worked through a male manager, such as Ostrowski, she found locally hired staff to be 'very supportive'.[73]

Channelled through intermediaries with a broad vision like Stéphane Ostrowski, the Saudi support came without censorship. The NWRC was thus a place not only of efficiency, but also of academic freedom. With this freedom, the foreign scientists supported by the NWRC were able to venture beyond the centre's core objective of houbara breeding. Williams found that Ostrowski 'promoted about everything' he and Tieleman did. Towards the business-minded Jacques Renaud, Ostrowski 'always portrayed what we did in a favourable light', said Williams. 'We had carte blanche essentially to do what we wanted.'[74] According to Ostrowski's vision, the NWRC was concerned with the very broad field of 'desert ecology', which covered all of Williams' and Tieleman's research. High-quality research outputs were more important on a scientific island of efficiency than considerations of religious sensitivities. Ostrowski was primarily interested in 'good science', which was published in 'good journals'. He gained confidence that the foreign researchers were delivering very quickly.[75]

The relationship between the NWRC and visiting scientists, such as Williams from America and Tieleman from the Netherlands, was not one-sided, but based on mutual interests. In return for funds and logistical support, Gulf-based managers like Ostrowski benefitted from the experience of scientists in the West. Williams helped Ostrowski gain a second doctoral degree in ecophysiology in 2006, adding to his previous doctorate in veterinary science.[76] For this second degree, Ostrowski and Williams researched the physiological adjustments of antelopes to desert environments. Their results received coverage by Fox News and *Science* magazine, demonstrating that international scientific networks could also connect Gulf-based scientists to international media networks. The two scientists found that when restricted in their access to food and water, gazelles shrank their livers and hearts. This allowed them to reduce their consumption of oxygen, which in turn enabled them to breathe less often and to reduce water loss. Williams commented in *Science* that 'the ability of the camel to survive the desert environment pales when compared to the sand gazelle'.[77]

In parallel to their work on sand gazelles, Joe Williams and Irene Tieleman carried out research into the evolutionary biology of desert birds. This research was part of Tieleman's doctoral studies at the University of Groningen, which she completed under Williams' supervision in 2002. The two scientists examined a hypothesis promulgated by researchers working on North American species that 'desert birds do not possess unique physiological adaptations to their environment'. The North American species were around 15,000 years old, and thus relatively young on an evolutionary time scale. In contrast to their colleagues, Williams and Tieleman suggested that some desert birds 'may have evolved physiological mechanisms' that promote low metabolic rates and low rates of evaporative water loss.[78]

The support and resources of one of the Gulf's scientific islands of efficiency made a long-term partnership with the NWRC attractive for Williams and Tieleman. After the completion of Tieleman's doctorate in 2002, the two scientists continued studying the evolution of desert birds in Saudi Arabia. In 2005, Williams and Tieleman published an article in the journal *BioScience* entitled 'Physiological Adaptation in Desert Birds'. Continuing Tieleman's doctoral work, the article called 'into question the idea that birds have not evolved unique physiological adaptations to desert environments'. In contrast to Saudi textbooks, which taught adaptation, but rejected Darwin's theory in general, Williams and Tieleman called adaptation 'the centrepiece of Darwinian evolution'. The scientists criticised previous authors for suggesting that birds had a pre-existing physiology suitable for the desert environment. In contrast this suggestion of 'pre-adaptation', Williams and Tieleman argued that birds had evolved such a physiology after moving into the desert.[79]

The implications of the Saudi-sponsored transnational research on birds were not restricted to Arabian animals and environments. Their research in the kingdom led the ornithologists to make general statements about the evolution of desert birds. In their *BioScience* article, Williams and Tieleman suggested that the Arabian deserts contained 'strong selection pressures on the physiological attributes of animals'. Such attributes included 'adjustments' that minimise energy expenditure or water loss and that enhance tolerance of high temperatures. Based on their own research, the scientists claimed that 'selection has reduced oxygen consumption' and total evaporative water loss among desert birds. Williams and Tieleman hoped that their evidence was sufficient for students to 'question the notion that birds do not possess physiological adaptations to the desert environment'.[80]

Despite the contradictions between Williams' and Tieleman's research and Saudi textbooks, the research did not create tensions with the leadership of the NCWCD. This lack of tension partly resulted from a lack of knowledge among the commission's directors and the wider Gulf's public about the details of the scientific work. In this case, a *dis*connection, rather than a connection, facilitated work on evolution in the Gulf. 'Although proud of the scientific prestige of the NWRC, and its positive outcomes for science in Saudi Arabia, the Saudi administration was at best mildly interested or even for a number of important decision makers not informed', said Ostrowski. The French scientist added that the 'evolutionary questions we developed in the publications never triggered any problems with our Saudi sponsors'. He speculated that the Saudi princes funding the NWRC 'were not reading our papers or did not want to see the potentially conflicting outcomes of these discussions'. Ostrowski added that 'since the results of our research work were paradoxically less propagated nationally then on the international arena we never faced any counter-productive pressure from our Saudi sponsors'.[81]

Ignorance or denial of the link between small and large evolutionary change further contributed to the lack of public controversy about the research at the NWRC. The disconnection between different parts of evolutionary theory hence made research into small-scale evolution easier. Joe Williams doubted that many Saudis 'fully understand the concept of evolution'. The zoologist added that Saudis 'warm up' quickly to the concept of 'microevolutionary' adjustments of animals to changes in their environment. However, when confronted with 'macroevolutionary change' relating to humans, Saudis preferred the Qur'anic account of creation. 'For any Saudi scientist', the American claimed, 'the Qur'an trumps any notion of human evolution, always.'[82]

The disconnection between evolutionary biology and the Gulf's public was, however, not total. While mainly confined to international journals, the research by Williams and Tieleman also fed into publications by Saudi institutions. Although using English rather than Arabic, such publications were more readily available for readers in the Gulf than American journals like *BioScience* were. In 2010, a British ornithologist named Michael Jennings published his *Atlas of the Breeding Birds of Arabia*. This *Atlas* appeared as a monographic volume within the series *Fauna of Arabia*, which MEPA and the NCWCD had sponsored for decades. Jennings's *Atlas* relied on numerous surveys undertaken with funding from the NCWCD since 1984. In the *Atlas*, Jennings stated that 'birds have evolved a number of characteristics' that helped them to conserve water, thermoregulate and avoid heat. Referring to Tieleman and Williams's research on hoopoe larks, Jennings found it 'likely' that the physical and behavioural adaptations of hoopoe larks were also repeated in other species.[83]

## Climate change

Despite the importance of the work undertaken at the NWRC, much of this work – and the status of the NWRC as an island of efficiency – ended in the late 2000s. This largely resulted from a policy of Saudisation and an associated change in the research climate. The replacement of foreign with Saudi staff was part of larger efforts by the Gulf governments to increase the number of nationals in the workforce. From the 1950s onwards, when the Gulf monarchies established their first universities, they considered foreign faculty members 'transitory'. They were mostly employed on renewable contracts, whereas nationals were allocated tenured or tenure-track positions.[84] Nationals also received scholarships in order to gain doctorates from universities abroad and replace foreign professors after their return. After a decline in oil revenues and continuing population growth during the 1980s, 'Saudisation', 'Emiratisation', 'Omanisation' and other 'Gulfisation' efforts became more visible in most sectors of the economy. They aimed at reducing unemployment and the potential for social upheaval. At a time of relative austerity after the oil boom of the 1970s and early 1980s, they also sought to keep capital in the domestic economy.[85]

Among the Gulf monarchies, Saudi Arabia, Bahrain and Oman, which had the least oil and gas revenue per citizen, faced the most acute problems of unemployment. They thus made the greatest efforts at increasing employment among their nationals. In contrast, Qatar, the United Arab Emirates and Kuwait, with smaller citizen populations, were able to maintain higher rent-to-citizen ratios. Qatar, the richest country, made the least systematic efforts to nationalise employment.[86] In the 2000s, the Saudi monarch still had enough funds to establish a few new scientific islands of efficiency staffed by highly paid foreign experts, such as the King Abdullah University of Science and Technology. However, most of the new islands were located in Qatar and the United Arab Emirates – Qatar Foundation's Education City and New York University Abu Dhabi being among the most prominent examples. In Saudi Arabia, nationalisation affected institutions such as the NCWCD, which were of lower royal priority than KAUST.

The Saudisation policies targeted most sectors of the economy, but were most effective in the public sector, including the NCWCD. Perhaps distracted by an increasing number of other commitments, the senior princes did not shield the commission from the budgetary stagnation and Saudisation that affected other government agencies during the 1990s. The pressure on the NCWCD's wildlife centres to create jobs for Saudis thus increased without a parallel increase in budgets. Consequently, the NCWCD hired many Saudis on low salaries,[87] which made foreign experts comparatively expensive. 'They only accept salaries close to what they would earn in their own countries', wrote Abdulaziz Abuzinada.[88] Stagnating budgets meant that as the NCWCD hired (often less qualified) Saudis on lower pay, it reduced the number of highly paid foreign scientists.

In the context of Saudisation and budgetary stagnation, the link between Jacques Renaud's company Reneco, which managed the NWRC, and the NCWCD, was severed. Abuzinada claimed that Renaud had 'reached a high degree of frustration' after years of debates and frictions about budgets during the 1990s and early 2000s. At some point, Renaud proposed a new budget for the NWRC to Abuzinada, which he described as 'realistic'. In this proposal, Renaud requested another 3 million riyals (about $800,000) on top of an expected budget of 6 million riyals. Abuzinada rejected this proposal, and Renaud thus withdrew as consultant, ending the contract between the NCWCD and Reneco. Exhausted by the disagreements with Renaud, Abuzinada thought he had reached the end of a 'nightmare'. At the same time, he 'felt for the first time the weight of responsibility, as there was no one among the Saudi specialists in the commission who was able to replace the consultant's experts'.[89]

A rapid Saudisation of the NWRC followed the sudden withdrawal of Renaud, threatening the centre's status as an island of efficiency in the Gulf's public sector. With a reduced overall budget for the NCWCD, Abuzinada found it impossible to pay the salaries of the foreign specialists, although some of them had agreed to renounce parts of their benefits. The sudden departure of Renaud created further financial and administrative problems that required an increase rather than a reduction in staff. Responding to the government's Saudisation goals, the NCWCD thus advertised low-paying positions for Saudi graduates of universities and secondary schools. Although Abuzinada found the candidates not to be of 'the required standard', he hired ten of them to work for the centre. In addition, the NWRC gained a new Saudi director general, an ornithologist named Mohammed Shobrak, in 2005. The

Saudi botanist Abdul-Rahman Khoja, who had previously liaised between the centre and the NCWCD's headquarters in Riyadh, became Shobrak's assistant.[90]

Combined with the lack of capacity building among Saudi nationals, the rapid Saudisation was detrimental to the climate of research. Within a few years, the connections with foreign scientists that had sustained research on evolutionary physiology suffered from disruptions. By 2012, working at the NWRC had become 'less easy and less attractive' for Irene Tieleman. The Dutch researcher especially complained about 'Saudisation and the lack of freedom' compared with the period from 1997 to 2007. She lamented that the Saudis 'had gotten rid of all the expatriates at the NWRC without having qualified Saudis to take their place'. A 'power play about who should be the new director' among the Saudis ensued. Mohammed Shobrak remained the NWRC's director general for only two years. In 2007, the Saudi primatologist Ahmed Boug replaced him. Since then, 'the bureaucracy has increased and knowledge and appreciation of science has decreased', said Tieleman. Ultimately, the Saudis 'pushed out' most of the French experts.[91]

Particularly disruptive for the transnational scientific networks was the departure of people, such as Stéphane Ostrowski, who had maintained connections between the NWRC and scientists based abroad, like Williams and Tieleman. After the end of the contract between Reneco and the NCWCD, Ostrowski quit the NWRC, as many expatriates had done before him. In early 2005, Ostrowski left Saudi Arabia 'because of the economic necessity of Saudisation' and his ambition to complete his second doctoral degree in Lyon. 'Unfortunately', he said, 'when I left we had no Saudis ready to pursue the work.' Joe Williams and Irene Tieleman continued to visit the NWRC a few times for additional fieldwork and then ended their collaboration.[92]

The connections between the Gulf research centres and international conservation organisations also provided scientists like Ostrowski with opportunities to find new employment. After gaining his second doctorate in 2006, Ostrowski joined the Wildlife Conservation Society in New York. However, in 2012, Ostrowksi still looked back positively at his time in Saudi Arabia and even imagined returning to a scientific island of efficiency. 'I often miss these wonderful times of intellectual prosperity and great scientific interactions', he said. 'Sometimes I dream that I have an unlimited funding resource and that I am given the opportunity to resurrect the ecophysiology lab at NWRC.' Ostrowski also imagined attracting 'a promising generation of young Saudis' in order to teach them about their 'wild creatures'.[93]

The worsening of the climate for expatriate researchers in Saudi Arabia also coincided with a change of leadership at the NCWRD. This change brought a renewed focus on breeding houbaras as a hunting resource, but less funding for wider research into desert ecology. In 2006, Abdulaziz Abuzinada retired as secretary general of the commission,[94] a position he had held for 20 years. Bandar bin Saud, a member of the extended royal family, replaced him. In contrast to Abuzinada, Prince Bandar was not a scientist but a hunter. He was mostly interested in research on houbara bustards, and to a small degree on oryx, but lacked interest in wider desert ecology. At about the same time, Saud Al-Faisal, the foreign minister and managing director of the commission, withdrew his support for the NCWCD. The NWRC thus lost the patronage of a senior member of the royal family – a loss for which Bandar bin Saud was not able to compensate.

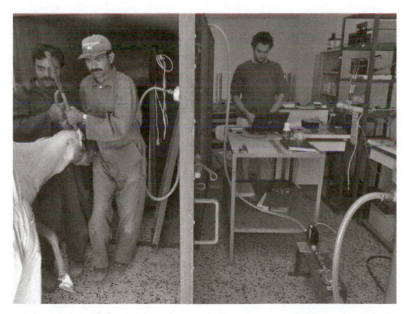

**Figure 4.2** Stéphane Ostrowski working in the physiology lab at the NWRC (courtesy of Catherine Tsagarakis)

**Figure 4.3**    Stéphane Ostrowski with a tranquilised Arabian oryx (courtesy of Eric Bedin)

As a self-proclaimed 'reformed hunter', Saud Al-Faisal had supported a 'modern and altruistic conservation orientation' of the commission, according to Ostrowski.[95] In addition, Saud Al-Faisal had donated several million riyals of his personal wealth to the Commission every year. 'That was a big stimulus', said Williams. However, when Prince Bandar took over the leadership, the funding for the NWRC 'essentially dried up', claimed Williams.[96]

A lack of qualified Saudi scientists exacerbated the departure of the French researchers. Foreign researchers like Williams and Tieleman did not integrate many Gulf nationals into their network of graduate students. A possible reason lay in a 'rentier mentality',[97] which assumes that income is less based on work and more on personal connections. Another reason was perhaps that there were still relatively few young Saudis interested in becoming zoologists. The NCWCD's recruitment of many Saudis based on patronage rather than merit further added to the difficulties. Tieleman claimed, 'we tried to include Saudis where we could'. Yet, she considered it 'not so easy to find Saudi students that would be interested in working with us, either in the field or in the lab'. The Dutch ornithologist also found many Saudis to be interested more in business and medicine rather than 'fundamental science'. Those Saudi students who joined the foreign researchers at the NWRC often entered the centre through family rather than academic connections. 'I was often not quite sure where people suddenly showed up from', said Tieleman, 'In the end it often turned out that they were the cousin of Abuzinada.'[98] Along with fast nationalisation, rather than a slower and gradual process – as in the case of Saudi Aramco – a clientelist rentier bureaucracy thus took over the former island of efficiency.

As the NCWCD's recruitment of many Saudis was not based on merit and scientific potential, their commitment to work in the field often did not match those of the foreign scientists. The foreign scientists, on the other hand, were ambitious and interested in science enough to accept the restrictions on lifestyle in Saudi Arabia. This gap between nationals and foreigners was a further obstacle in sustaining the networks between researchers in the Gulf and abroad in the absence of intermediaries like Ostrowski. The lack of commitment to biological fieldwork perhaps was related to a negative image that manual jobs in general had in the Gulf monarchies. After the oil boom of the 1970s, these jobs were associated with low-paid expatriates from Asian and African countries. However, even before the 1970s, manual workers, including artisans, had a low social status in the eyes of many Bedouin, and social interactions with them, including marriages, were restricted.[99]

Furthermore, in a rentier state, many Saudis entering the lower ranks of a governmental agency, like the NCWCD, perhaps assumed that their principal duty was being available in their offices during working hours (*dawām*).[100] This assumption was very different from the expectations that Joe Williams had toward his students. 'I tried, and tried, and tried to get Saudi students to work with us', Williams claimed. Sometimes, he was successful, and Saudi students joined his team 'for a few weeks'. 'Then', however, 'the rigour of having to get up at 4:30, to get out in the field and work all day just was too much. So, they quit.' Williams considered this lack of commitment among his Saudi students a 'big disappointment'.[101] Williams thus ended up relying on, and teaching, Western graduate students like Tieleman, rather than the Saudis.

The clientelism of the NCWCD was not entirely disruptive of transnational research, however. A few Saudi scientists became important partners of foreign ornithologists and key members in the Gulf's transnational scientific networks. Innovative biological research in Arabia, including research on the sensitive topic of evolution, was thus not exclusively the domain of foreigners. One of the most important Saudi ornithologists was Mohammed Shobrak, who served as director general of the National Wildlife Research Center between 2005 and 2007. He was a member of the second generation of biologists trained in the Gulf monarchies, the generation after Abdulaziz Abuzinada. In 1987, Shobrak gained a bachelor's degree in zoology from King Saud University. During this time, the Saudi government was building up the NCWCD, and Abuzinada was looking for Saudi biology students to join the commission. At the suggestion of one of his professors at the university, Shobrak met Abuzinada and immediately received a position as assistant biologist at the NWRC.[102] Very impressed by Shobrak, Abuzinada described him as 'an example of ambition, seriousness and willingness to work under all conditions'.[103]

Professional ambition and networks with foreign institutions were themselves important in enabling a Saudi zoologist like Shobrak to become an equal partner of Western scientists. After hiring him, Abuzinada used his connections to help Shobrak complete his education in Britain despite the NCWCD's budgetary constraints. Shobrak's 'willingness to develop his professional and academic potentials' impressed Abuzinada. During the 1990s, however, the budgets of the NCWCD as well as other governmental agencies were tighter because of lower oil prices and the expenditure on the Gulf War. The NCWCD was thus unable to provide Shobrak with a scholarship for graduate studies. However, through his 'personal contacts', he persuaded the British Council to grant Shobrak a scholarship.[104] Shobrak still needed to make personal sacrifices in order to support himself and his family during long periods of fieldwork in Saudi Arabia, which the British Council did not fund. 'I sold even my carpet to continue my PhD', he said. In 1996, he completed a doctoral thesis at the University of Glasgow on the ecology of the lappet-faced vulture in Saudi Arabia. Later, Shobrak fought a rentier mentality among his own students, telling them, 'if you want to do something, you have to be patient and work hard'.[105]

The contact with foreign scientists further inspired and encouraged Shobrak in his work. Shobrak called the NCWCD his 'mother', from which he learned 'all the issues about wildlife'. He specifically acknowledged to have learned 'something' from all the foreign experts at the NWRC, 'in life or in science'. Shobrak found Joe Williams' work habits particularly inspiring. 'Seeing an old man working hard' convinced Shobrak that he 'should be better also'.[106] The ornithologist Phil Seddon also claimed to have 'recognised the potential of people such as Mohammed Shobrak early on' and supported them. It was 'a great source of pleasure' to see how far Shobrak had come, he said.[107]

Despite his intensive contact with Western biologists, Shobrak appropriated the theory of evolution only selectively and reconciled it with his religiosity. 'I believe in evolution', the Saudi zoologist said, but without 'bringing the human in the issue'. He did not see this belief as contradicting his 'religious part' and described himself as a 'Muslim'. He thus claimed the widespread notion of harmony between science and religion in the Muslim world. Shobrak also remembered that he had a student at the University of Bath, whom he encouraged to 'work on' evolution in one of his papers. However, Shobrak himself avoided writing about evolution in general terms. Rather, he was trying to look at the issue 'from a conservation point' of view and tried to 'understand how the birds adapted' to high temperatures and other climatic conditions.[108] Accepting an adaptationist microevolution, but not macroevolution, Shobrak co-operated with Joe Williams and Irene Tieleman in researching the adjustments of birds to the desert climate. Shobrak initially observed that larks used the burrows of lizards as a refuge from the desert environment. With the help of Williams and Tieleman, Shobrak turned this initial observation into a research article. The three ornithologists argued that the use of burrows provided 'significant savings to the water economy of Hoopoe Larks', concluding that this behaviour is 'potentially important to their survival'.[109]

The Saudisation of the NWRC did not end the collaboration between Shobrak and his foreign partners. While the NWRC lost the patronage of Saud Al-Faisal, the Saudi higher education system benefitted from increased government spending with the rise in oil prices after 2007. The government established new universities in many provinces, thus offering positions for researchers in many fields, including biology. This was part of a wider growth in the number of universities in countries of the Gulf Cooperation Council from 40 in the 1990s to 117 in the 2000s.[110] After serving as director general of the NWRC for two years, Shobrak became head of the biology department at the recently established Taif University in 2007. This new academic position forced him to keep publishing and collaborating with his foreign colleagues.[111]

Like other biologists in the Gulf, Shobrak and Williams were not only interested in evolution, but also in the future survival of desert species. In the context of rising global concerns about climate change, Shobrak and his American colleague investigated increasing temperatures as a particular threat to Arabian animals. While he did not believe in human evolution, Shobrak did believe that 'climate change is happening'. He thus became interested in 'how animals can deal with it' and how scientists could 'support the wildlife'.[112] Williams and Shobrak's research on the effects of climate change, however, required the consideration of local cultural sensitivities. In 2011, a lecture by Williams on the impact of climate change on desert animals in Saudi Arabia sparked a small controversy. The Saudi Biological Society invited Williams to give this lecture at Taif University. As a public university in a city, Taif University probably offered less protection than the more isolated institutions, such as the NWRC or KAUST. After displaying a picture of a dog in America during his lecture, Williams showed the photo of a dog living in the Saudi climate. This dog wore the traditional headgear in Saudi Arabia, a scarf with a black cord. Williams said that he 'was trying to be funny but at the same time make a point'. He claimed that a 'Saudi national' had given him the picture, which made him assume that it was 'safe'.[113]

Despite Williams's humoristic intentions, the picture of a Saudi dog caused a negative reaction. Although he had visited the kingdom numerous times, Williams had not fully adjusted to the cultural sensitivities of audiences in the Gulf. Representative of the multinational composition of populations in the Gulf, his audience not only consisted of Saudis, but also of other Arabs. According to Williams, three Egyptians in the audience 'strongly objected' and 'got up'. Williams interpreted this reaction as an expression of 'distrust and dislike of American people in general'. He claimed that he apologised immediately and the rest of the audience 'clapped'. After the lecture, he stayed at his hotel and worked in the Mahazat as-Sayd Protected Area, not experiencing 'any sort of problems all week'. Then, however, he found the story being 'blown up' by the newspapers.[114] Perhaps responding to anti-American sentiments among its readers, the Gulf media reported that Williams' audience had 'considered the picture as an insult to the Saudi national identity'. According to the media, the university management had 'told the professor to stop the lecture and leave the campus and the Kingdom'.[115]

As in research on evolution, the disconnection between scientific journals and the wider Gulf public was paradoxically advantageous to the dissemination of Williams' results among academics. Regardless of the negative publicity in the Gulf newspapers, the main organ of the Saudi Biological Society, the *Saudi Journal of Biological Sciences*, still published an article based on Williams' lecture. Despite the word 'Saudi' in its name, this journal had a transnational character, which perhaps distanced it from the local media. A journal of King Saud University, it was published by the Dutch publisher Elsevier and had an international advisory board that consisted of scientists from outside the Gulf. Williams' article, which was co-authored with Mohammed Shobrak, warned that global warming was 'occurring at an alarming rate', affecting animals the world over. Yet, global warming particularly threatened the desert animals of Saudi Arabia, because they already faced high air temperatures, while having little access to water.[116]

In their article, Williams and Shobrak also warned that the speed of climate change exceeded the time necessary for evolutionary adaptations. The scientists estimated that Saudi Arabia would experience an increase in air temperature of three to five degrees Celsius over the course of the twenty-first century. They thus predicted that local animals 'will not have sufficient time to make evolutionary adjustments'. Adding to these concerns, Williams and his colleague found that the body temperatures of the Arabian oryx, spiny-tailed lizards, lappet-faced vultures and hoopoe larks sometimes already approached 'the upper lethal limit' of 47 degrees Celsius. The biologists thus warned that 'that global warming will have a detrimental impact' on Arabian wildlife. Given the urgency of this situation, Williams and Shobrak called on 'scientists in Saudi Arabia to study the impact' of increasing air temperatures on animals in the kingdom further.[117]

Research on climate change was not restricted to individual endeavours by Williams and Shobrak. Members of the third generation of professional biologists from the Gulf monarchies did indeed take up the ornithologists' call for more studies. This was the generation of Shobrak's own students, including a Saudi ornithologist named Monif Alrashidi. Like Shobrak, Alrashidi benefitted from an international education and was able to combine local environmental knowledge with global ecological expertise. Born to herders of sheep and camels in the desert near al-Qasim in central Saudi Arabia, Alrashidi exposed himself early to animal life. Over the years, he developed an interest in the behaviour of desert animals.[118] Between 1995 and 2004, Alrashidi studied biology at King Abdulaziz University, gaining a master's degree with a thesis on the ecology and protection of falcons. Subsequently, he enrolled at the University of Bath in England and pursued doctoral research on the ecology and conservation of the Kentish plover in Saudi Arabia.[119] While his formal supervisors worked in Bath, Mohammed Shobrak, who was then at Taif University, informally supervised Alrashidi during his fieldwork in the kingdom.[120]

As Alrashidi had received much of his education in Saudi Arabia, it is not surprising that he, like many scientists in the Gulf, was a creationist. As in the case of Shobrak and others, Alrashidi's acceptance of the theory of evolution was only partial, focusing on adaptations and ignoring speciation. He described himself as a 'Muslim who believes that there is a Creator and that all organisms on our planet were brought about by the Almighty Creator'. He also believed that there is 'evolution occurring in the behaviour and anatomy of living organisms'. This evolution was aimed at the 'adaption' of organisms to their natural environment and its changes. He gave the example of people possessing large lungs and living at high altitudes, which he considered 'an adaptation to the environment due to the lack of oxygen in high areas'. However, Alrashidi did 'not believe in Darwin's theory of evolution' in the sense that 'one organism can become another'. Keeping an open mind, however, he claimed that his conviction did not prevent him from 'studying' Darwin's theory.[121]

Alrashidi's creationism also produced certain tensions with Western scientists, but they were not detrimental to his education and work. The networks of science thus allowed for transcultural compromises. Alrashidi's rejection of Darwinism caused discussions with his supervisors at the University of Bath. At the beginning of his studies, his Hungarian supervisor at Bath, Tamás Székely, asked him to study not only the behaviour of birds but also their 'evolution based on fossils in order to find links with living organisms'. Alrashidi 'refused', however, and explained to his supervisor that he 'did not believe in the theory of evolution'. Székely then sent his Saudi graduate student to Matthew

Wills, another scientist at the university, for another 'lengthy discussion'. In the end, Alrashidi asked Wills, 'If there is really evolution, as you say, do you think that after millions of years, man will have wings, because I simply want to fly?' Wills then 'smiled' and allowed him to do what he 'wanted'. In 2010, the Saudi student completed a doctoral thesis that covered the behaviour and conservation of living birds, but excluded fossils.[122]

Even a creationist scientist from the Gulf like Alrashidi was, however, unable to keep evolution out of his work entirely. Through his collaboration with evolutionists, the theory still entered some of his publications. In 2011, Alrashidi published an article together with his supervisor Tamás Székely and Mohammed Shobrak, in which they analysed parental cooperation among Kentish plovers. The scientists assumed that 'selection should favour biparental care if it substantially improves the survival of the offspring'. Alrashidi and his colleagues found that Kentish plovers spent more time incubating in exposed nests than in nests shaded by bushes. This led the scientists to conclude that solar radiation necessitates parental cooperation in the Kentish plover. The scientists further stated that 'parental care often coevolves with mating strategies'. They thus conjectured that in different environments 'mating systems and parental care can diversify over evolutionary time'.[123]

Subsequently, Monif Alrashidi joined Mohammed Shobrak in studying the effects of climate change on birds in Saudi Arabia. Among the evidence for climatic change, he cited the 'dramatic rise in temperature in only fifty years', 'the rapid melting of ice in the polar regions' and 'the rise of sea water levels'.[124] In 2012, Alrashidi, Shobrak and his former supervisor Székely published a case study of the Kentish plover on the Farasan Islands in the *Saudi Journal of Biological Sciences*. In their study, Alrashidi and his colleagues warned that global warming would result in an accelerating rise in sea levels. A rise by one metre would flood 5 per cent of the coastal areas of the Farasan Islands, thus threatening the habitat of the Kentish plover.[125]

## Conclusions

From the 1980s onwards, the governments of Saudi Arabia and Abu Dhabi invested considerable resources in the breeding and reintroduction of houbara bustards. As the houbara was the most important prey in falconry, its reintroduction aimed at saving a royal pastime and an important part of the national heritage. At the same time, collaboration with international organisations in conserving houbaras provided Gulf sheikhs with 'environmental credentials'.[126] During the late 1980s, Saudi Arabia and Abu Dhabi founded two major research centres, the National Wildlife Research Center and the National Avian Research Center. While focusing on houbara breeding, the scientists at these centres broadened their research over the years to cover desert ecology in general. In the course of this research, they discovered and described important physiological adaptions of animals to the heat and aridity of Arabian environments. Several researchers considered these adaptations the result of long processes of evolution and thus warned that animals would not have the necessary time to adapt to further global warming during the twenty-first century.

The connection between senior princes and foreign entrepreneurs and scientists provided the Gulf's research centres with the characteristics of islands of efficiency within the Gulf's public sectors. These characteristics included patronage by a senior player in the regime, a clear mandate regarding output, managerial autonomy and the merit-based employment of highly paid foreign experts. Through competitive hiring, scientists and staff often came to the research centres through connections spanning Arabia itself and reaching scientists in different continents. As islands of efficiency, the centres were insulated not only from the interference of the larger clientelist bureaucracy, but also from the wider Gulf public. The resulting lack of awareness of the details of the scientific work among the Gulf authorities and public prevented controversy and thus provided freedom for research, including research on the sensitive topics of evolution and climate change. When, however, the patronage by a senior prince ended, the NWRC became vulnerable to nationalisation efforts, clientelism and budget cuts, thus causing the departure of foreign experts and disruptions to the research.

While the ornithologists and veterinarians in the Gulf discovered important evolutionary adaptations in desert animals, they mostly focused on contemporary birds and mammals and ignored fossils. This largely resulted from the mandate of the research centres to reintroduce contemporary species. However, in the case of the ornithologist Monif Alrashidi, the avoidance of fossils also resulted from a rejection of Darwinian evolution. Nevertheless, as the next chapter will demonstrate, a number of palaeontologists in the Gulf spent most of their time searching for, describing and interpreting fossils, including those of primates. In these fossils, microevolutionary adaptations and adjustments were less easy to separate from questions of speciation and macroevolutionary change. I will thus examine to what extent these fossils provoked new discussions about the sensitive topics of human origins and evolution.

# CHAPTER 5

# MISSING LINKS

In December 1991, an unusual exhibit opened at the Natural History Museum in London. It presented fossils from Abu Dhabi, an area that had rarely featured in international museums until then. The Natural History Museum thus connected palaeontology in the Gulf to audiences in the global city of London. At the same time, the exhibit itself resulted from the Gulf's connections with Western business and academia. The Abu Dhabi Company for Onshore Oil Operations (ADCO), a joint venture between the Abu Dhabi National Oil Company and Western petroleum firms, sponsored the exhibit. Representatives of the Western shareholders of ADCO, which included British Petroleum, Shell, Total, Exxon and Mobil, all attended the opening. What was unveiled in their presence were fossils, which a team of researchers from the NHM and Yale University had found in Abu Dhabi. Also part of the exhibit was a video entitled *Hot Fossils from Abu Dhabi*, featuring the British broadcaster David Attenborough.[1]

The exhibit and the video drew a stark contrast between the contemporary climate of Abu Dhabi and that of 8 million years ago. David Attenborough described the contemporary emirate as 'a wasteland wilderness of terrifying extremes and one of the harshest environments in the world'. Invoking the evolutionary concept of 'adaptation', he explained that 'only the most adaptable' living creatures 'can survive in the sparse vegetation and scorching summer temperatures'. However, the fossils displayed at the exhibit demonstrated that 8 million years ago, Abu Dhabi had a similar climate to contemporary East Africa, including a wet and dry season. 'Scrubby vegetation covered the now barren vegetation, and rivers teemed with fish, bathing hippos and hungry crocodiles', Attenborough narrated.[2]

In 1992, shortly after the exhibit at the Natural History Museum, the Abu Dhabi Company for Onshore Oil Operations produced a similar documentary, which also included concepts associated with evolutionary theory. The documentary's very title was *Abu Dhabi – The Missing Link*. Abu Dhabi television aired it several times,[3] and in 2005 it became part of a permanent exhibition at the newly established Environment Agency – Abu Dhabi,[4] which also took over the National Avian Research Center. In the documentary's opening sequence, the presenter promised to tell 'the story of how a team of scientific detectives found a missing link in an age-old mystery in Abu Dhabi'.[5]

The phrase 'missing link' was a popular reference to fossils that exemplified a transition from one species to another, especially in human evolution. Indicating the idea of a connection between humans and the rest of the animal kingdom, the phrase had been widely used even before the publication of Darwin's book *On the Origin of Species* in 1859. It originated in Christian as well as Muslim ideas about hierarchical structures and orders of nature that Enlightenment thinkers called the 'Great Chain of Being'. In the twentieth century, the missing link gradually became the 'ultimate prize in palaeoanthropology'. Many evolutionary biologists rejected the notion of a single missing link connecting humans to apes and rather spoke about common descent. Despite this criticism, the term remained a celebrated icon of human evolution. As such, even critical palaeoanthropologists used it, at least when it came to promoting their own findings.[6]

Despite the strong resonance of the term 'missing link', human evolution only remained a connotation in ADCO's documentary. Rather than referring to an intermediate form between humans and other apes, the 'missing link' meant a piece in the puzzle of how certain animals, like elephants, came to be found in both Asia and Africa. The documentary's presenter stated that because Arabia's climate 8 million years ago was similar to East Africa today, it was able to sustain animals that required a lot of vegetation, such as elephants. Hence, these animals were able to move from Africa, through Arabia and into Asia, and vice-versa. The documentary announced that 'the very important fossils from Abu Dhabi, the very first from the whole of Arabia, complete the missing link story of the movement of animals from Africa'.[7]

This chapter is not restricted to the coverage of evolution in biological exhibits and films produced in the Gulf. It is rather devoted to the search for another kind of 'missing link'. I investigate the connections and disconnections that made the discovery and presentation of fossils in the Gulf monarchies possible. Taking the exhibit at the Natural History Museum as one of my cases, I am especially interested in the connections between palaeontologists, museums, governments and oil companies. I am mostly concentrating on the period from the 1970s onwards, when some of the most important palaeontological discoveries in the Gulf, and especially of primates, occurred. This is also the period in which palaeobiologists in the West started using the fossil record to contribute to the theory of evolution in increasingly sophisticated ways. Drawing on the ever-growing computational power, palaeobiologists developed influential models of macroevolution and mass extinction.[8] I will thus also ask to what extent palaeontologists in the Gulf not only used the 'missing link' as an iconic phrase, but also made claims about the origin of species.

The previous chapters have already described the important role that connections with foreign scientists and institutions played in Arabian botany, conservation and ornithology. In palaeontology, the role of transnational and transregional networks was even more important. Whereas Gulf universities established departments of general biology, zoology and botany soon after their own foundation, most of them lacked departments of palaeontology. Similarly, the Gulf governments established museums of natural history, important institutions for palaeontology, rather late. In 1985, the Oman Natural History Museum opened in Muscat as one of the first institutions of its kind in the Gulf monarchies.[9] In 2014, the Qatar Museums Authority was still in the process of establishing a Qatar Natural History Museum. Indicative of the lack of qualified nationals, foreigners played leading roles in these museums. The British ornithologist Michael Gallagher was the leading advisor at the Oman Natural History Museum.[10] The German editor of the series *Fauna of Arabia*, Friedhelm Krupp, served as the founding director of the Qatar Natural History Museum.[11]

As Gulf universities and museums entered the field of palaeobiology rather late, this chapter pays special attention to scientists' networks with other institutions, including the Gulf tourism authorities, oil companies and Western universities and museums. The 'palaeontological patronage' provided by these institutions was crucial. Academic expertise and sheer luck were never enough for the discovery of fossils. Connections with resourceful institutions were important in order to gain concessions for excavations, equipment, supplies and permission to move fossils out of the country.[12] These connections made any palaeontological discovery not just a scientific, but also an economic and political affair. The search for fossils was also an especially political issue in the Gulf monarchies. Unlike other branches of biology, palaeontology also became the subject of major debates at universities and in the media, especially in the more open emirate of Dubai. These debates flared up as the publications by the prominent Turkish creationist Harun Yahya entered the Gulf markets and academia in the 2000s.

Amidst the politicisation of palaeontology, two kinds of 'fossil economies'[13] sustained palaeontological research in the Gulf. A fossil macro-economy, based on the production and consumption of oil and gas, fuelled and funded the general development of infrastructure and institutions, which palaeontologists and other scientists used. A fossil micro-economy consisted of smaller networks of exchange, in which scientists, oil companies and other institutions traded remains of organisms, logistical and bureaucratic support, scientific expertise, prestige and environmental credentials. Because of the abundance and broad distribution of microfossils, micropalaeontology was particularly useful for petroleum geologists in analysing sedimentary rocks. For that reason, micropalaeontology allied itself with the oil industry and won the support of oil firms for the broader discipline of palaeontology.[14] In the Gulf, the presence of many oil companies and geological surveys counterbalanced the lack of palaeontological university departments in the region. They supported major excavations and sensitive discoveries, including those of hominoids.

## Fossil economies

Geologists working for Western oil companies were some of the first scientists who collected and described fossils in the Gulf monarchies. As the companies' main aims were commercial rather than scientific, most of the reports of these geologists remained unpublished.[15] In the 1930s, for instance, oil geologists discovered two teeth from a mastodon in Saudi Arabia, which entered the field records of Aramco, but not the academic literature.[16] However, in a few cases geologists published their findings, attracting the attention of palaeontologists elsewhere. A cooperative mapping project of the Arabian peninsula between the US Geological Survey and Aramco resulted in a paper printed in 1966, which reported fossil vertebrates.[17]

One year later, in 1967, geologists working for Royal Dutch Shell in Abu Dhabi published an article in the journal *Palaeogeography, Palaeoclimatology, Palaeoecology*. Like other international journals in which findings from the Gulf monarchies appeared, this periodical was unaffected by censorship, but was also inaccessible for broad sections of Gulf societies. A disconnection between scientific work and the wider Gulf public thus existed in parallel with the transnational connections between geologists and palaeontologists. In their article, the Shell geologists described the discovery of plant-root structures in the Baynunah area of western Abu Dhabi. These structures, which were associated with low-lying swamp-like environments, provided one of the first indications of a less arid climate in Arabia 8 million years ago. To the geologists' surprise, they had found these root structures on a dry desert hill named Jabal Barakah. The geologists also discovered a tooth that a palaeontologist in Munich subsequently identified as belonging to a large mastodon from the Pliocene. Together with the root structures, the mastodon suggested the presence of considerable vegetation in Abu Dhabi several million years ago.[18]

During the late 1960s, Harold McClure, a micropalaeontologist working for Aramco, also began collecting fossil mammals as a hobby. By 1973, he had accumulated over 100 specimens, including a rhinoceros jaw. Although he had acquired some skill in recovering larger fossil bones, he was looking for specialist collaborators in order to study his fossils further. He first approached the Smithsonian Institution and other American museums, but they expressed little interest in his discoveries. Then, in 1973, he contacted Roger Hamilton, the head of the fossil mammal section at the Natural History Museum in London, requesting the identification of specimens and information on the relevant literature. In return, McClure shipped his material to the museum at Aramco's expense. Aramco also sponsored visas for a visit by Roger Hamilton and his colleague Peter Whybrow to the kingdom and arranged logistical and some financial support. In 1974, the two scientists collected fossils from two sites and shipped them to London. Their visas, however, forced them to restrict their field work to two weeks.[19]

The scientists from the Natural History Museum considered the collection very important for the study of mammalian evolution during the Miocene. In a note from 1975, Hamilton wrote that 'the Saudi Arabian locality is of great significance as it lies on the probable migration route between southern Asia and Africa and may therefore provide indications of the origin and stage of evolution of many mammalian groups'.[20] Besides the discovery of antelopes, Hamilton stated that the discovery of hominoids identified with specimens from East Africa clearly enhanced the importance of the site 'at least emotionally'. Hamilton and Whybrow thus hoped to return to Saudi Arabia for further collections.[21]

While the Natural History Museum scientists attempted to gain visas for a second visit to Saudi Arabia, the museum's director forbade any announcement of the discovery of hominoids. He feared that such an announcement would attract scientists from American institutions to the sites before the Natural History Museum palaeontologists could return. Furthermore, McClure and his British collaborators had collected the fossils and shipped them to London without permission from the Saudi government. The director feared that 'complications might arise as a result of this'. Until 'the situation regarding the Saudi Arabian authorities had been clarified' and a second expedition had been completed, Hamilton and Whybrow were not allowed to publish their findings.[22] This second time, however, Aramco refused to sponsor an expedition. While the company recognised that another trip 'could yield material of wider scientific interest', it stated that 'such more detailed work lies outside the scope of our practical interest'.[23]

Over several years, Roger Hamilton attempted to gain visas from Saudi institutions, without mentioning the discovery of hominoids. In these attempts, the Natural History Museum sought to trade expertise and prestige for support and specimens. In a letter to the University of Petroleum and Minerals in Dhahran from 1976, the Natural History Museum offered to 'supply expertise in collecting and research'. Hamilton also wrote that their cooperation would 'lend considerable prestige to both parties and would help the University to enter a field of geology and biology that is new and interesting to its staff and students'. Finally, the museum offered to instruct members of the university in extracting, preparing and identifying materials.[24] In another letter to

King Saud University from 1977, the Natural History Museum offered training and research facilities in return for transportation and logistical support in the field.[25]

In negotiating with Gulf institutions, the Natural History Museum also tried to keep at least some of the collected fossils. In his letter to the University of Petroleum and Minerals, Hamilton expected that 'an equitable division of the collection could be agreed'.[26] In another letter to the British Council in Riyadh, the NHM's keeper of palaeontology wrote that while his museum undertook research, 'the prime function of the staff is to conserve, curate and enhance the National Collections'. Research collaboration thus 'depends on an equitable division of the material collected'. In return for facilities and expertise offered to Saudi institutions, the museum's collections 'should be augmented for the benefit of international science'. The keeper thus considered 'one of the biggest problems … the complete banning by the Saudi authorities of the export of any specimens'.[27]

While the negotiations with the Saudi universities continued, competition with other international researchers forced the Natural History Museum to publish the results of their initial collection and disclose the discovery of the hominoids. Harold McClure warned that the university in Dhahran 'could not care less about vertebrate fossils, being mainly concerned as the name implies with petroleum and minerals, and negotiations could drag on *ad infinitum*'.[28] King Saud University expressed an interest in sponsoring an expedition, but still required more information in 1977.[29] Around the same time, Roger Hamilton received news that palaeoanthropologists from California and France were arranging collections in the area of Dhahran. Although he did not feel that 'the risks of some reaction from Saudi Arabia have been removed', Hamilton thus asked the NHM's director for permission to publish his findings.[30] In 1978, he, Whybrow, McClure and a colleague named Peter Andrews published two articles on the primates and the remaining fauna in the journal *Nature*, challenging an earlier view that Arabia had been barren of vertebrate fossils from the Miocene. The scientists also concluded that the primates had affinities with Miocene dryopithecines from East Africa, which meant that the area of Dhahran was close to the migration routes between Africa and Asia.[31]

While the *Nature* articles probably lent prestige to Hamilton and his collaborators, their removal of the fossils without permission damaged their relationship with the Saudi authorities. Peter Andrews recognised that the collection 'was taken out of the country in a sufficiently doubtful way for it to be something of an embarrassment'. He and his colleagues from the Natural History Museum also realised that they 'would be unlikely ever to work there again'.[32] Following the publication of the articles, Abdullah Masry, the director of the Saudi Department of Antiquities and Museums, tried to sack McClure from Aramco. In a letter to Peter Whybrow, Masry also complained about Aramco's 'high handed methods in disposing of fossils'. He demanded the return of the collection, before allowing any further work by the Natural History Museum in Saudi Arabia.[33] Roger Hamilton died in 1979, and in 1987, Peter Whybrow was still waiting for a decision by the Saudi High Council for Antiquities on a further excavation.[34]

Unable to return to Saudi Arabia, Peter Whybrow continued his work in Qatar and Abu Dhabi, which had a similar geology, but a more open visa regime for British citizens. This work, which began in 1979, again relied on connections with oil companies. While the Natural History Museum allocated the – at the time – considerable sum of £1,000 to his exploration, Whybrow received field support, travel facilities and advice from Qatar Petroleum and the Abu Dhabi Company for Onshore Oil Operations. He described the area as a harsh environment, in which 'goats sustain their life by eating immigration forms' at the UAE border. Yet, he found traces of a great abundance of life in the geological past. Stopping at Jabal Barakah, Whybrow found rods that he considered the fossil casts of mangrove roots. In an article written together with Harold McClure, Whybrow concluded that eastern Arabia had not been arid during the Miocene. Instead, the two scientists considered the climate near the coast to have been tropical or subtropical, with seasonal and perhaps monsoonal rainfall. This allowed for mangrove vegetation on the coasts and grasslands with shallow, vegetation-fringed rivers further inland.[35] Subsequently, Peter Whybrow and Peter Andrews engaged in a joint project with an Egyptian geologist at Qatar University on 'Eocene and Miocene Stratigraphy'. This project resulted in a chapter in a volume on *Primate Evolution* published by Cambridge University Press. This chapter provided an additional geological context for the discovery of the hominoids in Saudi Arabia.[36]

Independent from the Natural History Museum's explorations, the UAE's Department of Antiquities and Tourism in Al Ain started sponsoring excavations in the fossil-rich Baynunah region of Abu Dhabi. In March 1983, a team of five archaeologists surveyed parts of the coast of Abu Dhabi. This team not only located archaeological sites, but also discovered a large collection of fossils from the Miocene, including a hippopotamus jaw, elephant bones and fossilised plants. One of the members of the survey was an Iraqi archaeologist named Walid Yasin, who had joined the department a decade earlier as part of cultural and scientific co-operation between the Arab Gulf states.[37] Yasin recognised the significance of the fossils and searched for specialists to study them. He thus contacted Hans-Peter Uerpmann, a German archaeologist who was studying animal bones from a Bronze Age site in the Eastern Region of Abu Dhabi. Uerpmann promised to assist Yasin in finding an appropriate specialist.[38]

Through an element of chance, Uerpmann then connected the Department of Antiquities and Tourism to an anthropologist at Yale University named Andrew Hill. On a visit to New York, Uerpmann met Hill, who was studying human evolution in Africa at the time.[39] 'By chance', Hill said, Uerpmann informed him about the discovery in Abu Dhabi.[40] Hill was interested, and the Department of Antiquities and Tourism subsequently invited him to help evaluate the fossils. 'Luckily', as Yasin put it, Hill planned to fly to Pakistan around the same time and was able to stop in Abu Dhabi for a few days.[41] Hill subsequently connected with Whybrow, establishing a palaeontological network that stretched from Abu Dhabi to England and Connecticut. In 1986, the two palaeontologists submitted a joint report to the Department of Antiquities and Tourism in Al Ain, recommending that further work be carried out on the fossil sites. Two years later, the department invited Whybrow and Hill to organise an expedition, which resulted in a project on Miocene fauna and flora in Abu Dhabi. The aims of this project included locating new sites, interpreting the sediments, and recovering and dating fossils.[42]

Although the research by Whybrow and Hill related to evolution, this appears not to have been a point of controversy. This lack of controversy partly resulted from a separation of human and animal evolution in the minds of some scientists from the Gulf region. Walid Yasin remembered that evolution as such 'wasn't discussed' in the project. One of the reasons for the lack of controversy was, according to Yasin, that the project did not explicitly address human evolution, but 'palaeontology on animals'. The bones under investigation were 6 to 8 million years old and did not belong to 'humans', but, at most, to 'pre-humans, much pre-humans'. The Department of Antiquities and Tourism in Al Ain was thus able to cover the scientists' hotel and transport expenses without any problems.[43]

As their research was not controversial in Abu Dhabi, Whybrow and Hill neither engaged in self-censorship nor turned their interests away from evolution. Rather, Hill continued his work on human origins in East Africa alongside his research on Eastern Arabia's Miocene fossils. In 1988, Hill and a colleague published an article in the *American Journal of Physical Anthropology* entitled 'Origin of the Hominidae: The Record of African Large Hominoid Evolution'. Hill and his colleague wrote that hominids 'probably originated in Africa' between 14 and 4 million years ago. Subsequently, the authors discussed the factors that led to the development of the bipedal gait as the distinguishing factor between humans and other great apes. Hill and his colleague suggested an environmental change from tropical forest to grassland as a possible factor behind this development.[44]

Hill did not restrict his work on evolution to the African context, but, in collaboration with Whybrow and Yasin, also analysed Abu Dhabi's Miocene fossils with reference to evolution. In 1990, the three scientists chose the *Journal of Human Evolution* as a platform for publishing their discovery of primate and other animal fossils in Abu Dhabi in 1989.[45] In 1991, Hill and his colleagues also published an account of their findings in *Tribulus*, the bulletin of the Emirates Natural History Group (ENHG). Although this bulletin appeared inside Abu Dhabi rather than abroad, the scientists did not refrain from referring to evolutionary theory. On the contrary, Whybrow, Hill and Yasin even started their article with a quotation by the French-Italian evolutionary biologist Léon Croizat: 'Life and Earth have evolved together.'[46]

Although Whybrow, Hill and Yasin were studying evolution, they continued to receive support from local oil companies. *Tribulus*, the local medium in which the scientists published their findings, itself enjoyed such support. This was due to the strong personal connections between the oil companies and natural history groups in the Gulf. As the company scientists spent a lot of time studying the present and pre-historic environments from which the oil originated, they developed an interest in the natural history of the region. The founder of the Emirates

Natural History Group, a Briton named John 'Bish' Brown, had previously worked for the Kuwait Oil Company and was an active member of Kuwait's Ahmadi Natural History Group. In the mid-1970s, Brown joined the main offshore oil company in Abu Dhabi, the Abu Dhabi Marine Operating Company (ADMA-OPCO). Following Brown, many of his colleagues at ADMA-OPCO also joined the ENHG after its foundation in 1976.[47] ADMA-OPCO, together with the United Arab Emirates University, then sponsored the production of the magazine *Tribulus*.[48] This was part of a wider trend among oil and gas companies to promote environmentalism as part of corporate social responsibility policies.[49] The Environment Friends Society in Abu Dhabi, for instance, awarded ADGAS (Abu Dhabi Gas Liquefaction Company) its golden honorary membership 'in recognition of its commitment to the cause of clean environment, and to thank its positive and effective contributions to the Society's activities'.[50]

Besides ADMA-OPCO, the Abu Dhabi Company for Onshore Oil Operations (ADCO) supported the Emirates Natural History Group and palaeobiology in Abu Dhabi. Expensive palaeontological work involving scientists from abroad thus depended on a strong connection with government agencies and private business. ADCO sponsored the exhibit in the Natural History Museum, the documentary *Abu Dhabi – The Missing Link*, and, through corporate membership, the ENHG. The general manager of ADCO, another Briton named Terry Adams, even served as vice-chair of the ENHG. Moreover, ADCO supported Peter Whybrow's expeditions to the Gulf for almost 20 years between the 1970s and 1990s. In their *Tribulus* article from 1991, Whybrow and his colleagues acknowledged that Terry Adams and his wife had been 'extraordinarily helpful in providing logistic support'.[51]

During the 1980s and 1990s, financial support from oil companies even increased. In 1990, Peter Whybrow submitted an 'Initiative for Funding Geological and Palaeontological Studies of Miocene Rocks' to the Abu Dhabi National Oil Company, applying for £125,000 over five years. This time, his proposal made clear that after study, conservation, identification and a temporary exhibition in London, the Natural History Museum would return all the fossil material to Al Ain.[52] In 1991, ADNOC's subsidiary ADCO started supporting Whybrow with a grant. Under Terry Adam's successors as general managers of ADCO this grant continued,[53] allowing for further work until 1997. Under the leadership of Whybrow and Hill, 35 scientists participated in the project, searching an area of around 200 square kilometres for fossils.[54]

ADCO's support for palaeontology was partly due to the personal interest of its general manager, Terry Adams, in the subject. Adams himself held a degree in geology and a doctorate in micropalaeontology.[55] An active member of the ENHG, Adams lectured on palaeontology and led the group on a trip to the various sites in the Baynunah region of Abu Dhabi in March 1991. 'Bish' Brown, the founder of the Group, reported on this trip in *Tribulus*: 'It was a wonderful weekend, but once again, we appeal to anyone who may have picked up interesting fossils to show them to the palaeontologists. They might just have "the missing link".'[56]

The support of the oil companies for palaeontology was not pure philanthropy, however. In the fossil economies sustaining Gulf palaeontology, the oil firms traded financial and logistical support for a green image. The personal interest of its chair Terry Adams aside, ADCO used its funding for natural history in order to raise its profile as an environmentally responsible company. In 1990, ADCO's magazine *Al Waha* ('the oasis') published an article entitled 'ADCO is a good citizen'. This article reported on a press conference in which Terry Adams launched an 'Environmental Awareness' campaign. Adams explained that 'although the primary focus was naturally the Company's own terminals, it is also ADCO's responsibility not to disturb the desert environment, but also actively to protect it ... Global consensus shows that depletion of the ozone layer and deforestation was a common concern'. In order to lend expertise and credibility to the 'Environmental Awareness' campaign, 'Bish' Brown and Peter Hellyer, then chair of the ENHG, participated in the press conference.[57]

The corporate environmentalism of the Abu Dhabi Company for Onshore Oil Operations was not restricted to the Gulf, but – like the firm's Western shareholders – was international. In 1991, it formed part of the company-sponsored exhibit at the Natural History Museum. In the video *Hot Fossils from Abu Dhabi*, David Attenborough held one of the fossils from Abu Dhabi in his hands. Attenborough warned about 'dramatic' climate changes over the next 200 years, comparing them to the changes 'that overtook Abu Dhabi over the last 8 million years'. Invoking the concept of adaptation, the presenter reminded the audience that Abu Dhabi's animals 'became extinct, because they couldn't adapt to those changes'. The video concluded, 'If we can continue to find links between our own activities and natural events, which may affect the global climate, we will be able to act today to prevent tomorrow's world becoming an eternal desert.'[58] Palaeontologists – like other biologists in the Gulf – thus found a way of making their subjects relevant to audiences and funders by combining their research into the distant evolutionary past with claims about contemporary climate change.

The network between foreign palaeontologists, government agencies and oil companies was relatively stable and continued to support research in Abu Dhabi throughout the 1990s. In 1995, ADCO co-sponsored the First International Conference on the Fossil Vertebrates of Arabia. The conference also enjoyed the patronage of Nahyan bin Mubarak Al Nahyan, the minister of higher education and scientific research and patron of the ENHG. The conference, like most sensitive biological research, took place in isolation from large sections of Gulf society. A brochure described the venue, the Dhafra Beach Hotel, as 'the perfect spot to run away for some peace and privacy. It's so seclusive – nobody will bother to look for you here.'[59] The conference resulted in a book on *Fossil Vertebrates of Arabia*, which Peter Whybrow and Andrew Hill edited and published through Yale University Press.[60]

This publication attracted further researchers to Abu Dhabi. Faysal Bibi, a young Palestinian palaeontologist with relatives in Abu Dhabi, read the book soon after its publication. Growing up in Lebanon, his way into palaeontology relied on chance and a wide family network common among the Palestinian diaspora. 'Many palaeontologists tell you they always loved dinosaurs and always wanted to study them', said Bibi,

'I don't have any of these stories. In fact, I didn't even know what palaeontology was.' He added, 'Growing up in the Middle East, it is not really something you're exposed to. We don't grow up going to natural history museums, even though I had, because I used to go to London a lot to visit my mother who lived there. But still, it wasn't really anything that I thought that you could study and become. It was really a process of chance.'[61]

Given that the teaching of palaeontology was limited in the Arab world, connections with Western higher education institutions were crucial to the formation of young Arab palaeontologists, like Bibi. It was thus in California and not in the Middle East that Bibi became a palaeontologist. Initially, he went to study at a community college in San Diego, where relatives of his lived. From there, he went to the University of California, Berkeley, 'not for any particular reason except that it offered a lot of things'. 'By chance', as he remembered it, he learned about palaeontology and 'decided to stick with it'. He started working at Berkeley's Human Evolution Research Center, whose director was Tim White, the discoverer of Ardi, the then oldest known human ancestor in Ethiopia. Bibi went with White on fieldwork to Ethiopia and 'learned from him and his work'.[62]

In 2002, family connections brought Bibi to the United Arab Emirates. As he was visiting relatives in Abu Dhabi, Bibi read Whybrow and Hill's book *Fossil Vertebrates of Arabia*, which Tim White had lent him for the journey. This book described sites that were close to the place where Bibi and his family were staying.[63] Standing on one of the fossil-rich hills, 'I pretty much lost my head in the excitement', Bibi said.[64] After emailing Hill and Whybrow, he learned that the two scientists had ended their fieldwork in the late 1990s. He thus saw the opportunity to lead his own excavation team. As Bibi's relatives happened to work in senior positions in the Abu Dhabi Public Works Department, its chair, Sultan bin Zayed, heard about Bibi's interest and offered patronage. 'Put together a team, I'll pay for it', the sheikh said.[65]

Like researchers on the Gulf's scientific islands of efficiency, Bibi was able to fund his research quickly through patronage rather than seeking funding in the more competitive and bureaucratic American environment. Bibi's previous fieldwork with the Berkeley palaeontologist Tim White in Ethiopia had provided him with knowledge of surveying, collecting, conserving and preparing fossils.[66] Yet, as a young man without a doctorate, he would have faced difficulties gaining funding from American bodies. With Sultan bin Zayed's support, Bibi was able to start his own excavation together with colleagues from Berkeley. The Abu Dhabi Public Works Department paid for plane tickets, hotel accommodation, a car, a cook and food. Recalling this generous and perhaps rather naïve support, Bibi stated, 'it was ridiculous. It was wonderful.'[67] He thus tried to replicate the approach used for the Ethiopian excavations led by Tim White. Looking back at his excavations, he said 'despite our inexperience at the time, I think we did okay'.[68]

While quicker and less bureaucratic than a relationship with the American National Science Foundation or other major Western funding bodies, patronage by an individual sheikh from the Gulf was also less reliable in the long-term. The Abu Dhabi Public Works Department was politically marginal and soon lost Sultan bin Zayed as chair.[69] In this context, Bibi, like other internationally mobile biologists in the Gulf, sought to keep his connections with Western institutions. As he sought a career in palaeontology, the economics of academia soon forced Bibi to interrupt his adventure in Abu Dhabi. The young palaeontologist realised that his fieldwork in the emirate would take a long time to result in publications. The fossils were fragile and fragmented and initially 'not in any kind of condition that can be studied.' Cleaning and reconstructing the fossils in a laboratory could have taken years. Meanwhile, 'you have little to show for your work' in terms of publications, he said. 'That's difficult sometimes to justify when you need . . . a job.' In 2004, Bibi thus began graduate studies at Yale University, analysing specimens of antelopes that were already available in museums. This work was more attractive to him at the time, as it permitted him to publish 'more regularly'.[70] Maintaining a connection to Abu Dhabi, however, he chose Andrew Hill as one of his dissertation advisors.[71]

In 2004, soon after his retirement from the Natural History Museum in London, Peter Whybrow died. After a few years, however, Faysal Bibi and Andrew Hill returned to Abu Dhabi together and reactivated Hill's palaeontological network. In 2006, they received an email from Mark Beech, an archaeologist from the newly established Abu Dhabi Authority for Culture and Heritage (Adach). Adach had absorbed the Department of Antiquities and Tourism in Al Ain and the Abu Dhabi Islands Archaeological Survey, which had collected a lot of fossils. With an invitation from Adach, Hill and Bibi met Mark Beech and Walid Yasin. 'We all connected and reconnected', said Bibi. Together, they started a project a new project on 'late Miocene fossils of Abu Dhabi', which aimed to survey and excavate further sites.[72]

The new research by Bibi and his colleagues in Abu Dhabi covered the evolution of mammals. One of their most spectacular fossils was an entire trackway belonging to ancient elephants. Scientists sponsored by the Abu Dhabi Islands Archaeological Survey and ADCO had first announced the discovery of this site in *Tribulus*, the bulletin of the ENHG, in 2003.[73] The trackway allowed for new insights into the evolution of elephant behaviour. In an article in *Biology Letters* from 2012, Bibi and his colleagues stated that 'the origin and evolution of social structure in this clade is virtually unknown'. Yet, the fossilised trackway in Abu Dhabi formed, according to the scientists, 'early evidence' for a complex social structure among the ancestors of modern elephants. On the fossilised trackway, the scientists identified the footsteps of a herd which could have been matriarchal, and a solitary individual which could have been male. This led the authors to speculate that the social behaviour of modern elephants dated back to the Miocene.[74]

The research by the palaeontologists in Abu Dhabi did not remain restricted to international academic journals, such as *Biology Letters*. Because the fossils found by Bibi and his colleagues were often spectacular, their discovery received extensive coverage in the Emirati press. This was in contrast to much research on modern plants and animals, which – apart from the release of oryx or houbaras into the wild – attracted less media attention. The palaeontologists were thus able to connect with the wider Gulf public in a faster way than many other biologists were. Some of the coverage of the palaeontological work

**Figure 5.1** Faysal Bibi (right) and fellow researchers at the elephant trackway site in 2011 (courtesy of Mathieu Schuster)

**Figure 5.2**   Faysal Bibi (left) and fellow researchers at the elephant trackway site in 2011 (courtesy of Mark Beech)

even included references to evolution. In 2010, the Abu Dhabi newspaper *The National* reported on the discovery of an 8-million-year-old crocodile skull in Abu Dhabi. According to newspaper, Hill and Bibi believed that this ancient crocodile was 'closely related to species living in the Nile River today' and its head was thus easily recognisable. The paper then quoted Bibi saying that 'reptiles evolve very slowly, so this is not surprising.'[75] Through these occasional references, evolution even entered Gulf newspapers as some of the most strictly controlled media.

Faysal Bibi also promoted palaeontology as an evolutionary discipline in public lectures. In 2008, he gave a lecture at the American University of Beirut, the institution that had been the site of one of the first controversies about Charles Darwin's work in the Middle East in 1882. Entitled 'Ancestors Discovered: Paleontology and the Evolution of Life', Bibi's lecture asked his audience, 'Where does the diversity of life come from?' He responded by highlighting major palaeontological discoveries during the twentieth century, including 'evidence for the evolution of birds from dinosaurs and the evolution of our own species from a more generalized ape ancestor'. He then described his own work on fossil mammals in Ethiopia and the United Arab Emirates. Bibi concluded that 'since the publication of Darwin's major work in 1859, paleontology has provided the strongest evidence in support of the theory of biological evolution'.[76] In contrast to the speech from 1882, Bibi's lecture remained uncontroversial. Between 2008 and 2010, he accepted invitations to give a further four public lectures at AUB. During the same period, he also gave two presentations in Abu Dhabi and Al Ain with the uncontentious titles 'New Discoveries from the Yale University and ADACH late Miocene Fossil Project' and 'Fossil Discoveries from Al Gharbiya.'[77]

## Monkeys from Oman and the Hejaz

Oil companies and antiquities authorities were not the only major sponsors of palaeontological research in the Gulf. The palaeontological fossil economies included ministries, which the Gulf governments established during the decades after World War II. In the early 1950s, the Saudi government founded the Directorate of Oil and Mining Affairs, which turned into the Ministry of Petroleum and Mineral Resources in 1960. In 1975, the Kuwaiti emir decreed the establishment of a Ministry of Oil and the nationalisation of the foreign oil company in his country.[78] Agencies such as these tried to establish national sovereignty over natural resources during an era of decolonisation. However, governmental agencies still co-operated with postcolonial Western geological surveys in order to benefit from global expertise in exploring Arabia's natural resources.

One of the main collaborators in the Gulf geological surveys was the French Bureau de recherches géologiques et minières (BRGM). Like Kew Gardens, BRGM continued colonial research in the Middle East. Established in 1959, the BRGM incorporated several colonial survey organisations, including the Bureau minier de la France d'outre-mer and the Bureau de recherches minières de l'Algérie. During an era of decolonisation, the BRGM, like other European organisations, reinvented itself as an actor in the 'development' of other countries. With the end of colonial spheres of interest, its network of consultants expanded beyond the former French Empire and reached the Gulf, which had long been under British and American influence. In 1964, BRGM signed a contract with the Saudi government to survey the kingdom's resources.[79] From the Saudi perspective, this French connection presumably served to counterbalance the strong American influence through Aramco and the mission of the United States Geological Survey. This balancing of foreign influences was probably part of a wider strategy of the Gulf governments to award contracts to organisations from more than one country.

Although primarily searching for oil and mineral deposits, the survey geologists often discovered fossils, which brought them into contact with palaeontologists. As in the case of the Shell geologists' work in Abu Dhabi, the unexpected discovery of a mastodon in a dry area resulted in a collaboration between geologists and palaeontologists. In 1976, two geologists from the Bureau de recherches géologiques et minières discovered the remains of a mastodon tusk in the Eastern Province of Saudi Arabia. This prompted a brief reconnaissance in the region in 1977, which Herbert Thomas and Sevket Sen led. Thomas and Sen were palaeontologists with a special interest in primates and other mammals, who worked for the Musée de l'Homme, one of the departments of the Muséum national d'Histoire naturelle in Paris. The BRGM sponsored their expedition under an agreement with Ghazi Sultan, the Saudi deputy minister of mineral resources. The reconnaissance led to further excavations by Thomas and Sen in collaboration with Jack Roger, a BRGM geologist, between 1977 and 1978.[80]

As in the case of the Department of Antiquities and Tourism in Al Ain, an archaeological authority also became interested in sponsoring these excavations. In 1980, the Department of Antiquities and Museums, which was part of the Saudi Ministry of Education, supported a Saudi–French palaeontological expedition led by Herbert Thomas. The department provided vehicles, field equipment and support staff. It also sponsored *Atlal: The Journal of Saudi Arabian Archaeology*, in which the scientists published their findings. A transnational network between geologists and palaeontologists further contributed to the success of the expedition. The BRGM geologist Jack Roger and Peter Whybrow of the Natural History Museum provided advice and technical assistance. Thanks to this support by institutions and experts, Herbert Thomas and Sevket Sen were able to record 66 species, including 27 mammals, from the Miocene.[81]

During his work in Saudi Arabia, Sevket Sen, like other biologists, found the topic of human evolution to be taboo. Foreign scientists and Gulf officials were thus able to sustain fossil economies even in the face of different intellectual commitments. During the late 1970s and early 1980s, Sen had many discussions and developed 'quite close relationships' with Saudi officials and scientists. He found that many Saudis understood the importance of collecting fossils, especially for dating rocks. However, he remembered that it was 'delicate' and even 'forbidden' to talk about human evolution or to state that 'primates may have some ancestral relationships with humans'. According to Sen, his Saudi colleagues were 'closed' towards such discussions, or did not 'understand evolutionary processes of any kind, in particular not the evolutionary process in primates towards hominids'.[82]

As the networks of the BRGM were transnational, the French palaeontologists were able to use these networks to move from one Gulf state to another. During the late 1980s, Herbert Thomas and Sevket Sen also became involved in excavations in Oman, again in conjunction with the BRGM. This involvement resulted from an agreement between the BRGM and the Ministry of Petroleum and Minerals in Muscat to create a geological map of Oman. In 1968, the French government had charged the BRGM with publishing the geological map of France.[83] This made it attractive for the Omani Ministry of Petroleum and Minerals to collaborate with the BRGM in order to also create a geological map of the sultanate.[84] The Omanis perhaps also favoured the French in order to counterbalance the influence of the many British experts in their country.

A French–Omani team of geologists was thus put together, and palaeontologists subsequently joined them. One of the principal members of this team was Jack Roger, who had previously worked with Thomas and Sen in Saudi Arabia. Roger's main Omani collaborator was Zaher Al-Sulaimani, a young geologist who had just joined the Ministry of Petroleum and Minerals. The ministry charged Al-Sulaimani with working alongside the French scientists in order to facilitate communications and to write reports in Arabic. Roger and Al-Sulaimani were interested in collecting and dating the fossils in the rocks that they were surveying, as the age of the fossils also indicated the age of the rocks. The French–Omani team was used to finding marine fossils, because Oman possessed long historical as well as contemporary coastlines. However, during their mapping expeditions in 1985 and 1986, they also found fossils that belonged to terrestrial mammals. Roger and Al-Sulaimani thus searched for specialists on mammal fossils to help them with identification and dating. As a French–Omani team, their first thought was the Muséum national d'Histoire naturelle in Paris.[85]

In Oman too, the connections between geologists, local authorities and Western academic institutions enabled important palaeontological discoveries. Jack Roger's main contact was Herbert Thomas, who, in the meantime, had left the Muséum national d'Histoire naturelle to take over a chair in palaeoanthropology and prehistory at the Collège de France. Seeing some of the Omani fossils, Thomas and Sevket Sen decided to join Roger and Al-Sulaimani in conducting fieldwork in Oman.[86] Between 1987 and 1989, the scientists led three palaeontological expeditions to Dhofar. They received funding from the BRGM and the Collège de France, while the Omani Ministry of Petroleum and Minerals provided logistical help. At two sites, the scientists excavated and screened a total of about 15 tons of sediments. The mammals they found included representatives of 11 species, including rodents, hyraxes and elephants.[87]

Other mammals aside, the most widely publicised discovery of the French–Omani expeditions was one of primates. Unusual for biological research in the Gulf, even the British general-audience magazine *New Scientist* covered the discovery. The fossils were about 36 million years old, making them the oldest primates found outside Africa. Thomas and his colleagues were able to date them with unusual accuracy, because of the abundant remains of other mammals at the site. That such primates would be found in Arabia turned out to be logical. The Arabian peninsula was the closest landmass to East Africa, the region of previous primate discoveries. In addition, only a shallow sea separated the peninsula from Africa 36 million years ago.[88]

Unlike his experiences in Saudi Arabia, Sen did not find evolutionary theory to be taboo in his conversations in Oman. Sen claimed that his Omani colleagues were 'more educated' and 'more open-minded' than the Saudis he had met. The Omanis identified themselves as Muslims, but for them 'palaeontology, evolution, relationships between the animals, and also between the primates and man' were 'not a problem', said Sen. He thus found that he had a 'good basis of understanding' and scientific 'exchange' with them.[89] Sen's counterpart Zaher Al-Sulaimani echoed these remarks. He did not remember evolution to have been 'an issue' in his collaboration with the French team.[90]

It is thus not surprising that the French–Omani team repeatedly referred to evolution in their publications. These publications were mainly placed in international academic journals that lacked wider audiences in the Gulf and hence the potential for creating controversy. In an article in *Geobios*, Herbert Thomas and his colleagues wrote that the 'state of evolution' of the rodents and elephants suggests that the Dhofar sites were older than a previous site at Fayum in Egypt.[91] In another article in *Comptes rendus de l'Académie des sciences*, the scientists classified their primates according to an evolutionary scheme. They stated that two of the primates were 'Anthropoids', which revealed 'affinities' with primates previously found in Egypt.[92]

During the 1990s and 2000s, Herbert Thomas also wrote books on human evolution that were aimed at general readers of French. In these books, he integrated his findings in Oman into a longer story of human origins. In 1994, he published a small book entitled *L'Homme avant l'Homme: le scénario des origines*. In a chapter on 'the ancestors of our ancestors', Thomas wrote that the primates discovered in Oman and Egypt were 'early ancestors of apes and hominids'. He added that 'they are the first to possess, like us, thirty-two teeth'. As monkeys who lived on trees, these primates 'probably evolved in the rainforests bordering rivers and deltas'.[93] Like Faysal Bibi's lecture in Beirut, this book remained uncontroversial. In 2001, Thomas published another book entitled *Les Primates, ancêtres de l'Homme*, in which he also discussed his findings in Oman.[94]

While Thomas' books enjoyed a popular readership in France, they were even less likely to reach Oman, where the reading knowledge of French was limited. Thus, the frequent disconnection between scientific publications and the wider Gulf public also applied to palaeontological work in Oman. However, *Oman 2013–2014*, a French travel guide, covered the research by Thomas and his colleagues. This guidebook could potentially have enabled French-speaking tourists to talk with the local population about the discoveries. The guide stated that Herbert Thomas found the 'remains of the oldest known anthropoid primate' in Oman. These primates were 2 or 3 million years older than the primates found at Fayum. Hence, the primates 'represent a new stage in the ongoing race for the "missing link," in which all palaeontologists are engaged'.[95]

Despite the importance of the discoveries for the study of primate evolution, the French–Omani palaeontological excavations went through a hiatus during the 1990s and early 2000s. This was, however, not due to any controversy about human evolution. The main reason was rather that the researchers had examined the sites containing mammals thoroughly and were unable to justify new excavations. In a funding environment characterised by short-term planning, it was difficult to make a 'business case' for explorations over longer periods. Apart from the logistics inside the sultanate, for which the Ministry of Petroleum and Minerals paid, French institutions covered the main costs of the expeditions. Sen considered it 'not possible' to find thousands of francs every year to work in the sultanate.[96] Moreover, the geological map of Oman, which had been the subject of the original contract between the Ministry of Petroleum and Minerals and the BRGM, appeared between 1991 and 1993.[97] The formal collaboration between the French and Omani scientists thus ended.

However, the Gulf's scientific links with different foreign countries meant that even if one team ended their work, another team was often able to pick up the research. During the 2000s, researchers from the University of Michigan, including the Jordanian palaeontologist Iyad Zalmout, continued the search for fossils in Arabia. Like Bibi, Zalmout had been unable to study palaeontology as such in the Arab world. Instead, he studied the allied discipline of geology, which was more widely available at Arab universities, because of its importance in searching for oil and other natural resources. In 1998, Zalmout completed a master's degree in geology at Yarmouk University in Jordan. Subsequently, he pursued postgraduate studies in geology at the University of Michigan under the supervision of the palaeontologist Philip Gingerich. In 2001, Zalmout published an article in *Science* together with his supervisor and colleagues from the Geological Survey of Pakistan. They described the hands and feet of two 47-million-year-old ancestors of modern whales. Recalling a common view, Zalmout and his colleagues stated that 'the origin of whales involved an evolutionary transition from land to sea'. Yet, the fossil from Pakistan allowed them to offer the more specific conclusion that 'whales evolved from early artiodactyls'.[98]

Like the French palaeontologists, the scientists from Michigan also worked transnationally. Parallel to their work on fossils from Pakistan,

Gingerich and Zalmout searched for fossil whales in the Hejaz, as they presumed the region to have been under the sea during the Eocene. Having had previous experience working with the Geological Survey of Pakistan, the Michigan team collaborated with the Saudi Geological Survey. The Saudi government had established this survey as an organisation attached to the Ministry of Petroleum and Mineral Resources in 1999, when the operations of the BRGM ended. The survey combined parts of the ministry and the missions of the BRGM and the United States Geological Survey in the kingdom.[99] The president of the survey from 2006 onwards was Zohair Nawab, a Saudi geologist who had studied in North America and collaborated with the BRGM since at least the 1980s.[100] Under his leadership, the survey explored Arabia for dinosaur fossils. During the course of this exploration, Yousry Attia, an Egyptian palaeontologist working for the survey, recommended the University of Michigan as a partner.[101] In this partnership, the Saudi Geological Survey exchanged access to fossil-rich sites for international palaeontological expertise.

Together with this transnational expertise, chance was an important factor behind the palaeontological discoveries in the Hejaz. In 2009, Iyad Zalmout started excavating fossils north of Jeddah together with a team from the Saudi Geological Survey. Instead of Eocene whales, the researchers found parts of a skull of a mammal that was much smaller and much younger. Not being able to identify the creature, Zalmout sent an email with pictures back to his lab in Michigan, announcing 'more fun stuff' and asking whether the pictures showed 'a monkey or a horse'. Gingerich replied that the fossil seemed to represent a Miocene ape.[102] For further preparation and analysis by an expert on fossil apes, Zalmout sent the fossil to Michigan.[103]

The aim of the fossil's analysis was to contribute to a scientific debate about the split between Old World monkeys and apes in evolutionary history. Old World monkeys, which include hamadryas baboons in modern Saudi Arabia, have tails and are able to jump and swing from tree branches. Apes, which include gorillas, chimpanzees and humans, are tail-less and tend to have an upright posture. Researchers like the team at Michigan assumed that Old World monkeys and apes shared a common ancestry. At some point, the two lineages diverged, one giving rise to Old World monkeys and the other giving rise to apes and humans. Palaeontologists had long debated when this this split had occurred. A previous estimate, made using DNA samples of living primates, put the split between 35 and 29 million years ago. The Hejazi fossil found by Iyad Zalmout and his Saudi colleagues promised to be very relevant to this debate. 'It is neither a monkey, nor an ape', said Zalmout, but 'an intermediate primate that tells you a story about Old World monkeys and apes'.[104]

The discovery by the University of Michigan and the Saudi Geological Survey subsequently formed a cover story of the journal *Nature* in 2010. Zalmout and his colleagues argued that the fossil represented a new species named *Saadanius hijazensis*. This Latinised Arabic name itself reflected the transnational nature of the find. It adhered to the Linnaean system, but also acknowledged the region of discovery, as it derived from *Saʿdān al-Ḥijāz* 'monkey of the Hejaz'. The specimen, which was around 28 million years old, was anatomically close to the divergence between the apes and Old World monkeys, leading the scientists to conclude that the divergence occurred between 29 and 24 million years ago.[105] Zalmout also claimed that his discovery emphasised that 'Afro-Arabia is the center of catarrhine origin and evolution'.[106] He added that 'if you find one primate there should be more there'.[107]

In addition to the peer-reviewed article, *Nature* also disseminated the discovery in formats attractive to broader audiences. An article in simpler language on the *Nature News* website suggested that the 'plateau above Mecca in Saudi Arabia may soon attract pilgrims of palaeontology'.[108] *Nature* produced a short video in English and Arabic with the title 'A piece in the monkey puzzle.' Freely available on *YouTube*'s Nature Video Channel, it made *Saadanius hijazensis* better known than most other discoveries in Arabian biology. By December 2012, *YouTube* had counted around 9,000 views of the English version and 4,000 views of the Arabic version. One of the scientists from Michigan stated that 'humans, apes and Old World monkeys all share a common ancestor', adding that their specimen was 'very close to when apes and Old World monkeys would have diverged from one another'. Zalmout's supervisor Gingerich thus called *Saadanius hijazensis* an 'important missing link'. He acknowledged that 'some people don't like that term', but he did, 'because it fills a gap in time, it fills a gap in form, and it helps tie things together that we weren't able to tie together convincingly before'.[109]

The leadership of the Saudi Geological Survey did not reject the discovery, but endorsed and even reconciled it with nationalism, if not religion. In the Arabic version of the *Nature* video, Zohair Nawab, the president of the Saudi Geological Survey, appeared and expressed national pride about the finding and his agency's participation. Probably aware of the sensitivity of the topic, Nawab avoided talking about the relationship between humans and apes. Instead, he gave an uncontroversial statement that the research provides 'evidence of the geographical distribution of the apes at a time when the Arabian peninsula and Africa were one continent'. Yet, he did not indicate that the finding or its implications were in conflict with religion. 'This discovery is a unique scientific breakthrough on the local and global levels', Nawab concluded, 'It will enrich the geological and historical record of the Kingdom of Saudi Arabia, God willing.'[110] According to Zalmout, Saudi officials were also willing to preserve the site and promote fossil tourism.[111] Revenues from tourists would then further expand the fossil micro-economies in the Gulf.

Although Zohair Nawab backed the discovery, its implications for human evolution were still sensitive. The *New York Times* and *BBC Arabic* covered the discovery, but newspapers in Saudi Arabia only reported on them briefly and without mentioning evolution. In 2012, two years after the publication in *Nature*, Saudi Aramco's magazine *al-Qāfilah* published a four-page story about *Saadanius hijazensis*. However, while *al-Qāfilah* included a picture and the names of all the Saudi researchers involved, it did not mention evolution or *Saadanius*'s relationship with humans. This was in contrast to the BBC article, which was entitled '*Saadanius hijazensis* could be the missing link in the evolution of apes and man'.[112] In the comments section of the Arabic version of the *Nature* video on *YouTube*, one user wrote, 'unfortunately, we in Saudi Arabia are the last to learn about this discovery and about biological evolution as a whole'.[113]

The cautious media coverage inside Saudi Arabia probably protected Iyad Zalmout and his Saudi colleagues from public outrage. In this case, censorship helped scientists, at least in the short term. Rather than having to leave the kingdom because of the sensitivity of his discovery, Zalmout even found employment at King Saud University's Mammals Research Chair. In this capacity, he was involved in sensitive research in a very different field. In 2014, he co-authored a study that showed that the

Middle East respiratory syndrome (MERS) coronavirus had been circulating among dromedary camels in Saudi Arabia since at least 1992. Finding no evidence of infection in domestic sheep and goats, Zalmout and his colleagues suggested that camels were responsible for transmitting the MERS coronavirus to humans.[114]

Because of the internet and differences in censorship between Gulf countries, some people in the region still learned about the relationship between *Saadanius hijazensis* and humans. A Lebanese journalist also published an article on the topic on the website of *Al Arabiya*. Although the Saudis owned the news channel, it broadcast from Dubai and thus benefitted from a more liberal publishing environment. In an unusual synthesis of evolution and Islam, *Al Arabiya's* article emphasised the proximity of *Saadanius hijazensis* to Islam's holiest site and the house of Ibrahim or Abraham, the first monotheist according to Muslim tradition. The article was entitled 'The earthly ancestor *Saadanius hijazensis* lived near Mecca: Saudis and Americans found the most important missing link'. The article began by stating, 'Since man came down from the trees millions of years ago and stood up on his feet as a distinction among land animals, he ponders the mystery in order to get to know his secrets.' One of these secrets was 'the most important link, a being that lived near Mecca'.[115]

## Debates in Dubai

Thanks to internet-based media, small public debates were able to take place about *Saadanius hijazensis*. Through *YouTube's* comments section, the English and Arabic versions of the *Nature* video allowed different concerns to surface. The remarks on the English version mainly criticised the use of the term 'missing link'. The best-rated comment said that 'the concept of a missing link is inherently flawed', because it suggested 'a fixed chain of species' that 'evolved at a precise time, suddenly changing into another species'. Instead, the comment argued for 'gradual, fluid change over millions of years'.[116] The comments on the Arabic version of the video, in contrast, had less to say about the term 'missing link'. Instead, they represented a spectrum of views on science and religion. One viewer named Atheism750 declared that 'scientific progress drives religions into a very narrow corner'. Another viewer disagreed, saying that 'the theory of evolution is not incompatible with religion'. 'Many scientists believe in religion and are also convinced by the theory of evolution', the viewer added.[117]

Al Arabiya's article about Saadanius hijazensis too revealed a plurality of views on evolution among Arabophone audiences. In the comments section below the article, readers shared different views. One stated that 'the evidence for evolution increases as time goes by'. Another commentator by the name of Abu Muhammad disagreed. 'I don't know why the modern scientists insist on linking man to monkeys', he said, 'These theories and studies are merely an extension of Darwin's stupid theory, which contradicts the simplest facts.' Abu Muhammad added, 'The origin of man, dear scientist, is Adam – peace be upon him. He is a man created by God Almighty.' Yet another commentator joked, 'Why do you make excavations and spend money? In Jordan, there is a tribe named Hijazin that has not evolved much during the last 35 million years.'[118]

These discussions were part of increasing debates between creationists and evolutionists that took place on the internet from the 1990s onwards. The work by Andrew Hill and Faysal Bibi in Abu Dhabi featured in these debates too. Although both scientists believed in evolution, their research even allowed for an opposite interpretation. Because the 8-million-year-old fossils in Abu Dhabi looked similar to the bones of modern animals, Hill and Bibi's research lent itself to the Institute for Creation Research, a Christian institution from Texas whose materials had been 'cloned' in Turkey since the 1980s.[119] In 2010, the institute's website quoted Bibi saying that his fossils were 'recognisable versions and ancestors of the animals we know today'. The institute took this statement to ask, 'After all these supposed millions of years, why have today's versions of these creatures not evolved beyond the boundaries of their basic recognisable forms?' The website then – like some biologists in the Gulf monarchies – made a distinction between microevolutionary adjustments and the macroevolutionary emergence of new species. 'No fossilised transitional forms have been found in Abu Dhabi or elsewhere, which is contrary to what the broad picture of macroevolution would predict', the website said. 'These elephant fossils are definitely elephants and not just another branch on the evolutionary tree.'[120]

While these debates took place on the internet, they initially did not interrupt Hill and Bibi's excavations. Bibi, in fact, found evolution to be rather uncontroversial during much of the 2000s. The sensitivity of the theory thus often led to avoidance of the topic rather than open controversies. Bibi partly explained the absence of controversy with the lack of young earth creationism in Islam in comparison with Christianity. In his experience, Muslims generally believed estimates of the Earth as billions of years old. Yet, while Bibi and his colleagues talked about 'animal evolution a little bit' in their public lectures in Abu Dhabi, they avoided human evolution. Instead, they simply stated that around 7 million years ago, 'there was a whole diversity of animals', with rivers flowing in the area. According to the palaeontologist, local members of the audience reconciled this view with scripture. A common response was 'Oh yes, in the Qur'an it's said this area was once a paradise.'[121] As in the case of *Al Arabiya*'s article on *Saadanius hijazensis*, religious readers in Arabia could still accommodate sensitive palaeontological discoveries.

Another reason for the lack of controversy about evolution was again a disconnection between scientific research and the wider public in the Gulf. As most primary and secondary schools did not teach evolution, many people lacked knowledge of the theory and thus did not display any particular hostility towards it. Bibi diagnosed a lack of 'scientific awareness' in the Arab world in general. While this lack made his work easier, Bibi also considered it a 'big problem'. 'Most people don't have first-hand experience with science' and 'don't learn much of it in school', the palaeontologist stated. Even in Beirut with its reputation as 'one of the most open and academic cities' in the Middle East, he often had to introduce palaeontology on a very basic level. He claimed that popular audiences often heard about evolution from him for 'the first time'. Bibi, however, argued that 'evolution needs to be taught and needs to be discussed. It's a sad reality that it's not controversial.' 'It might make things easier in particular situations', Bibi concluded, 'but it's a sad product of a sad reality.'[122]

However, despite the lack of controversy surrounding his work, Bibi still recognised that evolution was a 'sensitive issue'. He claimed that he did not avoid the issue, apart from human evolution. While claiming to avoid 'censorship', he acknowledged, however, that 'you are always aware of where a potential red line could be that you don't want to cross unnecessarily, because you don't want to offend people'. He added, 'we honestly worry about finding a, for example, a hominid, a human ancestor'. The chances of it are 'very small' and 'it would be an amazing scientific discovery', Bibi said. However, the palaeontologist did not know how it would be 'received socially in the region and in the Emirates'. It could 'be received with pride and with open arms, and promoted' or 'it might be a problem'.[123]

Palaeontology and evolution became more controversial during the course of the 2000s and 2010s. During this time, the Harun Yahya's Science Research Foundation contested the work by palaeontologists, like Faysal Bibi and Andrew Hill. The Turkish author had already started challenging Darwinism and promoting an Islamic creationism in the 1980s.[124] Like the palaeontologists, he used fossils as evidence. In 2007, the Science Research Foundation sent hundreds of lavishly produced copies of the *Atlas of Creation* to scientists and schools in Europe and America. The *Atlas* accepted that the Earth was billions of years old, but rejected evolutionary change. Pictures of fossils, accompanied by modern-day organisms looking strikingly similar, filled the 500-page *Atlas*. Based on these pictures, Yahya argued that living creatures are just like the creatures that lived in the fossil past. Hence, the theory of evolution was false. Donations reportedly covered the costs of the foundations' activities, which were estimated at millions of dollars.[125] A creationist fossil micro-economy thus emerged in parallel to, and in competition with, an evolutionist one.

In the 2000s, Harun Yahya's literature entered the Gulf monarchies. A very receptive place was the United Arab Emirates, and especially Dubai. The emirate, which also hosted the Saudi-owned *Al Arabiya* network, was an easy entry point for books due to its economic openness and relatively liberal censorship regime in the Middle East. At the same time, the rulers of the United Arab Emirates had strong religious commitments. The emirate thus became a focal point for debates about palaeontology and evolution. Already in 2003, Yahya's books were on display at the Sharjah World Book Fair.[126] In 2008, the Kuwaiti newspaper *al-Ra'y* published an article entitled 'Freeing the Secularists from the Planet of Apes'. The article nominated Harun Yahya for the Islamic Personality of the Year Award, which Sheikh Mohammed bin Rashid, the ruler of Dubai, sponsored. *Al-Ra'y* justified this nomination by stating that Harun Yahya 'is known for his books that reveal the falsity of the theory of evolution, which traces man's origin back to the great apes and the origin of life to the first cell'. Yahya also 'highlights the close connection between Darwinism and bloody philosophies'.[127]

Harun Yahya did not win the Islamic Personality Award that year, but his organisation still gained ground in Dubai. The ensuing debates did not take the form of a contest between science and religion, however, but rather of allegations of pseudo-science. Harun Yahya's organisation tried to use Western universities in the UAE and science itself to argue against evolution. In 2007, the *Khaleej Times* reported that the Science Research Foundation organised an exhibition on 'Living Fossils' in Dubai. An official from the foundation claimed to 'prove scientifically that living beings that lived billions of years ago were the same as they are now and that the theory of evolution is nothing but a farce'. The organisers planned to hold the exhibition first at the American University in Dubai and the American University of Sharjah, two private institutions founded in the 1990s. Subsequently, the foundation planned to display the fossils 'in various shopping malls in Dubai for the people'. Whether in universities or malls, the foundation claimed to reveal 'the truth' to students. 'They need to know that the theory of evolution is just a propaganda theory', the official said, 'and we will prove this with the help of science'.[128]

The push against the theory of evolution, and evolutionary interpretations of fossils, was successful in the United Arab Emirates. In what appears to be a response to creationist pressures, the UAE Ministry of Education removed evolution from the curricula of state schools in 2007. Previously, evolution had been included in the curriculum for grade 12, according to Abdul Qader Eisa from the Ministry of Education. Darwin's theory was supposed to familiarise pupils with the historical progress of science. At the same time, the lesson on evolution also included viewpoints of Islamic scholars and a section on science and religion. In 2007, however, Eisa announced the removal of the lesson from the curriculum the following academic year. The official explained that the new curriculum would include 'crucial topics such as the human systems, and recent advances in DNA technologies'. This required the replacement of evolutionary theory. This concern not to overburden the curriculum was a common justification for removing evolution from Arab education systems. In Lebanon, officials excluded evolution 'from the examinable baccalaureate curriculum to ostensibly lighten the curricular load'.[129] The UAE still allowed private schools to address evolution without rejecting creationism. After all, Eisa said, 'we belong to an Islamic country, and we believe that scientific evidence proves the existence of God and that He is the creator of everything.'[130]

Despite the removal of evolution from secondary schools, biologists at foreign universities in the United Arab Emirates continued to teach evolution, while allowing contrary views by students. This drew members of Western universities into debates about evolution. In the relatively open media environment of Dubai, scientists could voice their opinions even in newspapers, and thus publicly oppose Harun Yahya's initiatives. One of these scientists was Aaron Bartholomew, an assistant professor of marine biology at the American University of Sharjah. In 2007, *Gulf News* quoted Bartholomew in an article entitled 'Debating the origin of life'. He stated that 'overwhelming' experimental, fossil and theoretical evidence supported the theory of evolution. According to him, 'No one has scientifically been able to disprove the theory.' Bartholomew gave the examples of cells, which 'are complex because of millions of years of evolution'. At the American University of Sharjah, he taught evolution and biodiversity and incorporated evolution into other courses such as ecology, genetics and marine biology. Yet, Bartholomew did not impose the theory on his students: 'I have taught students that disagree with evolution who received an A.'[131]

Despite resistance by some biology professors, the Science Research Foundation continued to use fossils and science itself – rather than scripture – in order to propagate creationism in the United Arab Emirates. In January 2011, Cihat Gündoğdu, an official from the foundation, gave a presentation at Zayed University in Dubai, which accompanied an exhibition of fossils of a Miocene polar bear and other animals. The university's office of student life organised the presentation in cooperation with the Islamic Affairs and Charitable Activities Department in Dubai. In his presentation, Gündoğdu asked his audience, 'Can chance lead to complexity and a variety of species? What are the odds of a single protein forming by chance?' The official also used quotes and clips from various scientists to point to what he claimed were gaps in the theory of evolution. He argued that instead of evolving, species were simply created as they are. Displaying images of various fossil forms, Gündoğdu claimed that no transitional forms of species, such as half-starfish and half-fish, had ever been found.[132]

Serving as a platform for debate, the Abu Dhabi newspaper *The National* reported not only on Gündoğdu's presentation, but also covered the opposing view. Despite evolution's sensitivity, the more open media in the emirates again gave voice to evolutionists using fossil evidence. The newspaper invited John Burt, then assistant professor of biology at New York University Abu Dhabi, to comment. Burt, who did not participate in the event at Zayed University, acknowledged gaps in the fossil record. 'We don't have every possible transition for every possible species', he said, 'but this should come as no surprise given the rare conditions that are required for fossilisation of tissues.' Moreover, carbon dating of existing fossils 'show clear chronological sequences', which contradicts the creation of life 'all at once'. Finally, Burt stressed that 'the theory of evolution underpins all aspects of the biological sciences, from genetics to ecology, and is widely considered to be among the ten most important ideas in the history of the sciences'.[133]

While most debates about palaeontology and evolution remained in the more open media environments of the UAE, Harun Yahya also propagated his views in the more restricted Saudi media. In 2008, the Saudi newspaper *Arab News* published an interview with the Turkish writer under the title 'Harun Yahya: Win over Darwinism.' Yahya stated, 'First, we offered Darwinists around the world 100 million fossils, which prove that this world came into being as a result of God's creationism and not because of evolution.' The Turkish writer continued, 'Darwin wrote in his books that people have to find transitional forms to prove the theory of evolution, but nobody has been able to find a single transitional form.' Third, it is 'impossible for even a single protein', let alone a cell, 'to be formed by chance.' Fourth, Yahya claimed that 'the skulls that were displayed as evidence of evolution are fake'.[134]

Harun Yahya's arguments reinforced those of members of the Muslim Brotherhood that had long been influential in the Saudi media. In 2003 and 2007, around 40 years after the death of Sayyid Qutb, *Arab News* printed two articles by the Egyptian author. Qutb argued that 'no animal has the necessary qualities to evolve to the human status'.[135] He further stated that the 'discovery of genes and chromosomes – which Darwin did not know about – makes progress from one species to another impossible. Every cell carries genes that preserve the distinctive characteristics of every species, and make it inevitable that it stays as a separate species.' Qutb also described people who thought of Darwin's theory as 'scientifically indisputable' as 'deluded'. Instead, the Egyptian author argued that 'the only explanation' was that God 'made man and breathed of His spirit into him'.[136] Despite important discoveries, such as *Saadanius hijazensis*, foreign palaeontologists were not able to counter creationists, like Sayyid Qutb and Harun Yahya, in newspapers in Saudi Arabia as they were in the UAE. In the early twenty-first century, Saudi debates about evolution – as well as about other sensitive issues – thus took place mainly on the internet.

## Conclusions

In the Gulf, as elsewhere, palaeontological findings depended on a degree of luck. Discoveries were not only difficult to predict, but some of them even occurred while palaeontologists were looking for something else. This happened to the discoverers of *Saadanius hijazensis*, who had originally been looking for whales. However, fossil economies consisting of networks of exchange between scientists, oil firms and governmental institutions facilitated these discoveries. Geologists working for oil companies or ministries were among the first people who found fossils. In order to identify these fossils and date the surrounding rocks accordingly, they invited palaeontologists working at Western museums and universities to conduct fieldwork in Arabia. As their own funds were limited, the palaeontologists in turn relied on these networks with oil companies and governmental agencies. In exchange for expertise, prestige and environmental credentials, they often gained logistical, financial and bureaucratic support for further work.

The search for missing links, however, sometimes also depended on disconnections between foreign scientists and the local population. That large sections of Gulf societies lacked knowledge of the theory of evolution and palaeontology made it less controversial for palaeontologists to publish on evolution. Moreover, scientists often published their results in international academic journals. While many of these journals had a high impact factor measured in terms of citations by other scientists, they were rather inaccessible for broader sections of Gulf societies. The Saudi newspapers also avoided evolutionary references when reporting on discoveries such as *Saadanius hijazensis*. This led to much research on evolution going unnoticed, but also undisturbed. The 'missing link' between scientists and the population thus made research sometimes easier. In the 2000s and 2010s, however, Harun Yahya's

Science Research Foundation gained ground in the Gulf, using some of the same fossils discovered by palaeontologists as evidence against evolution. In the relatively open media environments of the United Arab Emirates, the Turkish creationists also entered into debates with evolutionists at universities, in newspapers and on the internet.

# CHAPTER 6

# RENTIER SCIENCE

Arabia was no 'empty quarter' for modern biology and allied scientific disciplines. Since at least the eighteenth century, the peninsula formed part of transnational research networks connecting plants, animals, humans and their institutions. Already during Carl Linnaeus's lifetime, Peter Forsskål was describing and classifying Arabian organisms according to his teacher's system. During the nineteenth and twentieth centuries, many European naturalists continued Forsskål's exploration of Arabia by making extensive use of local knowledge and assistance. From the 1950s onwards, botanists from Egypt, Britain and other countries also built up departments of botany at the newly established universities in the Gulf. Besides educating the first generations of indigenous professional biologists, these expatriates established herbaria and produced comprehensive descriptions of Arabian plants. From the 1980s onwards, botanists from Germany also cooperated with the young Gulf universities in order to create vegetation maps and investigate oasis agriculture.

Although it was arguably not as important for agriculture as botany, modern zoology also developed in the Gulf monarchies. In addition to university-based research, scientists studied Arabian fauna at wildlife research centres and reserves. After initially focusing on the breeding and reintroduction of endangered flagship species, like the oryx or houbaras, scientists subsequently broadened their research agendas to include broad sections of desert ecology. In parallel to this research on contemporary desert environments, palaeontologists investigated past fauna and flora. In collaboration with geological surveys, oil companies and archaeological agencies, they published discoveries in major international journals, such as *Nature*.

Much of the modern biological research undertaken in the Gulf benefitted from, and served, rentier states. The Gulf governments paid for this research with the rents they received through the export of oil and gas. The energy stored in the fossil remains of past creatures thus fuelled the study of their relatives and descendants. In return for governmental support, biological research served the goals of rentier states. By examining fossils, palaeontologists helped date rocks and assess them for the purposes of extracting oil and minerals. Botanists also sought to contribute to agricultural development and food security for the Gulf's populations. In addition, researchers working on contemporary organisms sought to contribute to the economic future of rentier states in a post-oil era. Studying the region's renewable natural resources, that is living plants, animals and humans, these researchers sought to contribute to the diversification of the economy. Producing knowledge, they also sought to contribute to the replacement of oil-based with knowledge-based economies. Biologists and governments thus formed a symbiotic relationship, creating a rentier science.

Rentier science helped the Gulf states control their environments and societies. By surveying, mapping, identifying and classifying the region's flora and fauna, scientists rendered environments legible. They thus created inventories and maps of the natural resources that the states could exploit and develop. This scientific knowledge then formed the basis for high-modernist forms of environmental engineering, like in the oasis of al-Ahsa' in Saudi Arabia.[1] Scientists and conservationists were also instrumental in establishing protected areas. While these areas served the reintroduction of certain species of charismatic megafauna, they excluded pastoral nomads and their livestock from these areas. Some nomads found new employment as rangers, which brought them under the patronage of the state. This form of environmentalism was thus a form of 'social control'[2] in the context of wider governmental efforts to settle Bedouin.

The high modernism of states like Saudi Arabia and Qatar[3] was itself based on an evolutionary logic. In places as different as Egypt and China, conceptions of evolution had shaped intellectual debates, politics and literature from the late nineteenth century onwards.[4] The 'dogma of development'[5] that characterised Saudi politics thus had roots in discussions about Darwin as well as long-standing British imperial notions of the 'improvement'[6] of the world. A common Arabic word for 'development', *taṭawwur*, also meant 'evolution'. While the Gulf governments did not attempt to create new species, they sought to 'develop' and 'improve' human, animal and plant life on the Arabian peninsula. The Saudi state thus founded its wildlife authority under the name of the National Commission for Wildlife Conservation and *Development*.

While using science for environmental and social control, the Gulf states also sought to control science itself, and especially science education. The region's 'late rentier states' were undemocratic, but responsive to certain sections of their populations, such as ulema.[7] The Gulf governments thus balanced high-modernist technocratic initiatives with conservative religious stances. In the case of biology, this meant that the government responded to popular opposition to the theory of evolution, and especially human evolution, among their religious scholars and wider populations. At various times and to different extents, the Gulf governments excluded the theory from teaching at state institutions. During different periods, Saudi biology textbooks denounced Darwin's theory and censors in Oman and Qatar[8] blacked or cut out sections from textbooks. While many biologists still taught aspects of evolution at universities, they were conscious of the sensitivity of the topic.

The state-controlled media also responded to the religious conservatism of populations in the Gulf and the sensitivity of certain areas of science. Occasionally, Gulf newspapers published attacks on evolution. The more liberal publishing environment of the United Arab Emirates also saw the emergence of small debates between creationists and evolutionists. However, most media in the Gulf avoided discussions of evolution, and especially human evolution. The newspapers reported on biological research, such as the discovery of *Saadanius hijazensis*, but shied away from discussing its evolutionary implications and notions of shared ancestry between humans and other primates.

In the elite field of scientific *research*, late rentier states still enjoyed more autonomy from wider society than in mass media and education. Many government ministers were also either technocrats themselves or members of an older generation with little formal education and knowledge of modern science. They thus supported scientific research, sometimes in a naïve way, as long as it broadly served the states' developmental aims. Unable or unwilling to manage research to the extent that they controlled curricula, senior government officials gave scientists considerable freedom in pursuing their work. Helpful for the pursuit of a broad spectrum of scientific activities was also the Gulf states' toleration and support of scientists from diverse national and religious backgrounds, including Americans and Christians.[9]

While education was restricted, research had enough autonomy to be innovative even in religiously and socially conservative environments. Palaeontological research, which appeared in major journals such as *Nature*, changed views of past Arabian climates, showing an abundance of subtropical life several million years ago. Palaeontologists also discovered transitional fossils or missing links in the evolution of humans and primates. Other biologists investigated the adaptations of contemporary organisms to the desert environments. A few scientists even discovered evolutionary origins of certain taxonomic groups, such as Omani wheat races. Although most of these discoveries first appeared in international academic journals, some also entered media aimed at non-specialists, like *Arabian Wildlife* or the magazines of local natural history groups.

In addition to being sensitive, biological research in the Gulf was even critical of state practices at various times. This demonstrates that the scientific connections in the modern Gulf – like the scholarly networks in the medieval Arabic lands[10] – allowed for agency among scientists in dealing with sheikhs and princes. Although an ideology of development led the Gulf governments to support science, a number of biologists criticised rapid development and modernisation for damaging environments. Other biologists warned against the effects of climate change on Arabian animals, a process through which the Gulf states contributed disproportionately through their production and consumption of hydrocarbons.

The development of innovative research on evolution and climate change demonstrates that neither religion nor authoritarianism were fundamental obstacles to biology in the Gulf. Given the sensitivity of evolution, this means that the region's conservative forms of Islam, including Wahhabism or Salafism, were even less of a barrier to uncontroversial work in disciplines like physics or chemistry. Science and religion were not always in 'harmony' in Islam,[11] but, as historians of Christian missions have argued,[12] science and religion were not fundamentally opposed to one another either. Many biologists in the Gulf were also able to reconcile their beliefs in evolution with faith, at least when examining nonhumans. The resource-rich and developmentalist rentier states, although undemocratic, enabled scientific research more than they obstructed it.

## Scientific clientelism

An important feature of rentier science was scientific clientelism, a form of 'segmented clientelism', or a system of 'rent-based clientelism in which vertical links dominate'.[13] The top level of this system consisted of members of governments and ruling families who patronised administrators of scientific institutions. Within their institutions or segments, these administrators in turn acted as patrons for lower-ranking officials, scientists and doctoral students. Saud Al-Faisal, for instance, patronised Abdulaziz Abu Zinada, the secretary general of the NCWCD, who in turn was a patron for Mohammed Shobrak. Like fiefdoms in the security sector, such as the Saudi National Guard, scientific institutions thus contained chains of intermediaries who channelled the state's oil revenues down.

While the Gulf's patronage networks made some parts of the public sector 'sluggish' and 'overstaffed', they also allowed the creation of 'islands of efficiency'.[14] Patronage from a few senior government figures protected certain organisations from the corruption, nepotism and patronage-based rather than merit-based employment that characterised other parts of the rentier bureaucracy. These organisations included the wildlife research centres under the patronage of Saud Al-Faisal and other princes. Highly qualified and highly paid foreign experts managed and staffed them between the 1980s and 1990s. Princely patrons gave the foreigners a clear mandate and strong support to reintroduce selected species. Yet, they also provided the experts with enough autonomy to follow their own research interests, even into the sensitive area of evolution.

What contributed to the efficiency of the Gulf's wildlife research centres was their distance from wider society. Although physically based in Arabia, they were located outside major cities and in or near reserves. The foreign scientists often received on-site housing, funds, research facilities and support staff. They also did not share many of the duties of scientists at the Gulf's public universities, such as teaching and writing Arabic textbooks. This allowed the scientific islands of efficiency to grow their research output rapidly in comparison with the larger public universities. Hired on a temporary basis and without the prospect of meaningful integration into Gulf society, the foreign scientists did not restrict their ambitions to a career in the Gulf. They thus concentrated on producing research articles for international journals rather than local media. If academics in the Arab world had to choose between publishing locally and perishing globally or publishing globally and perishing locally,[15] most expatriate biologists chose the latter.

Dependence on personal patronage also made the scientific islands of efficiency vulnerable to changes in personnel at the top of the Gulf's regimes. Because they were isolated from wider parts of Gulf bureaucracy and society, they lacked wider support networks once key patrons disappeared. The scientific islands of efficiency were thus also similar to private businesses that employed few Gulf nationals and had little interdependence with society. In austere times, the government protected the interests of civil servants and consumers, who were mostly nationals, rather than private businesses and research centres staffed by foreigners.[16] When the state cut budgets, it thus hit the National Wildlife Research Center harder than agencies that employed more Saudis. As the aging Saud Al-Faisal withdrew his personal support from the NWRC in the 2000s, the centre soon lost its characteristics as an island of efficiency. Upon the departure of the NWRC's French consultant manager, Saudi bureaucrats established direct control over the centre. During this process, the centre lost funding, managerial autonomy and many of its foreign experts. Having failed to educate enough Saudis, scientific productivity declined and the centre lost much of its international visibility and reputation.

While clientelism was the order of funding in rentier science, the Gulf's resources also sustained vertical links. In contrast to the segmented clientelism of other governmental agencies, which communicated vertically with patrons rather than laterally with other agencies,[17] scientific clientelism also relied on communication across institutions. Collaborating scientists were often located at different institutions, such as universities, research centres, oil companies or geological surveys. This made the scientific networks in the Gulf more similar to Western academic and Islamic social institutions than other parts of the Gulf bureaucracy. Islamic social institutions often consisted of horizontal networks between members of Middle Eastern middle classes rather than of vertical relations between wealthier and poorer people.[18]

The peer networks between scientists were also often more egalitarian than the patron–client links that provided the funding for research. The structure of scientific clientelism was thus more like flat networks than a pyramid. Although scientists had different professorial ranks and salaries and institutional affiliations of different prestige, many of them did not create a hierarchy of patron–client relations. The authors of papers often appeared in order of their scientific contribution rather than their academic or professional ranks. The articles by the American professor Joe Williams and his Dutch doctoral student Irene Tieleman, for instance, sometimes have Tieleman, and sometimes Williams as first author.[19] Similarly, the Saudi ornithologist Monif Alrashidi often appears before his supervisors in the list of authors.[20] Non-scientist patrons, like Saud Al-Faisal, often received special thanks in the acknowledgement sections, but so did peer reviewers and other colleagues.

The horizontal peer networks not only crossed institutions, but also countries. Rentier science thus reached beyond the territories of the Gulf states. In the case of Williams and Tieleman, scientists at the National Wildlife Research Center in Saudi Arabia co-operated with peers from Ohio State University and the University of Groningen. Some of these Western universities perhaps appeared in international academic rankings far above their collaborating institutions in Arabia. However, the Gulf's scientific islands of efficiency often provided initial funding for research on Arabian organisms in a faster way than scientists were able to gain from larger and more bureaucratic agencies in the West, such as the United States National Science Foundation. This created a science in transaction, in which Gulf institutions provided the funds, equipment and local environmental knowledge and Western researchers provided much of the theoretical and experimental expertise. The scientific knowledge was often equally the product of the contributions of actors inside and outside of the Gulf.

That the transnational networks of Gulf science were horizontal rather than vertical did not mean that the networks were completely loose, however. Instead, certain institutions acted as hubs that concentrated connections. Through their control of the oil resources, the ruling families formed financial hubs for scientific and other forms of clientelism.[21] Institutions in control of other natural resources for scientific researchers, such as specimens, were often located abroad. European centres of calculation, like the Kew Royal Botanic Gardens, for instance, formed a port of call for botanists from many developing countries, including those in the Gulf. Botanists in Riyadh, Kuwait, Doha and Muscat visited Kew or corresponded with its scientists in order to identify organisms, access literature and to gain feedback on manuscripts and international scientific credibility. In return for their expertise, scientists in Kew, Paris or Berlin, often received rare specimens and local knowledge. This natural and intellectual currency allowed Western institutions to remain centres of calculation in the postcolonial era, when their budgets came under frequent pressure.

The networks of rentier science not only transcended national borders, but also the academic profession. If high modernism produced a 'rule of experts' in the Gulf, this rule depended on laypersons.[22] Amateurs interested in conservation and natural history acted not only as patrons of science, as in the case of princes and oil company executives; they also played a range of supporting roles. Botany and wildlife researchers relied on the knowledge and company of Bedouin rangers during fieldwork. Drivers and pilots provided female researchers in particular with logistical support and mobility to carry out their work. Finally, self-trained amateurs, such as the British botanist Sheila Collenette, themselves collected plants, contributed to herbaria and published papers. Other non-professional scientists founded local organisations, including natural history societies, which hosted and supported visiting professionals.

All of these connections were important in enabling scientists in the Gulf to undertake biological research, including sensitive and critical research on evolution and climate change. However, in some cases, a missing social link was also important for the study of links between species. The sheikhs and princes often did not read the publications that the scientists produced under their sponsorship. Similarly, broad sections of the population had little access to articles in international academic journals, the main research output of the scientific islands of efficiency. While local newspapers covered some research, they also avoided some of its more sensitive implications. This lack of connection between the scientists and the public hindered the firm establishment of evolutionary biology in the state curricula. However, the lack of public knowledge about evolution also made biological research less controversial and easier in some instances.

That Gulf patrons did not closely monitor the research they sponsored, and the fact that many other networks were horizontal contributed to agency and academic freedom among scientists in the Gulf. Although the Gulf monarchies were undemocratic, censorship seems to have had little tangible effect on academic publications, compared with the newspapers. Despite discrimination against Shiite nationals in some Gulf countries, such as Saudi Arabia and Bahrain,[23] I have not come across persecution of scientists based on their religious, ideological or ethnic backgrounds. Far from homogeneous, the scientific networks of the Gulf displayed considerable diversity. Scientists from the Gulf countries hosted, and worked with, African, Asian, European and American peers of various religious backgrounds. The general 'tolerance and openness' towards non-Arabs and non-Muslims[24] in the Gulf monarchies was very strong in the field of science.

This agency and freedom of Gulf scientists did not mean, however, that all Gulf scientists were active creators of their networks, as the popular notion of professional 'networking' suggests. Without wanting to overemphasise contingency, chance also played a role in Gulf science. Luck was not only involved in some of the discoveries of fossils; sometimes, scientists seemed also to stumble upon pre-existing networks. Often, they only learned by chance about opportunities to engage in collaborative research. Finally, as actor-network theory predicts, non-humans played an active role in the wider research process. Plants and animals themselves and the physical infrastructure, including cars, helicopters and laboratories, contributed to the outcome of scientific investigations. The Gulf's scientific networks thus comprised not only social networks, but also human–animal relations and technical and financial networks, including fossil macro- and micro-economies.

## Transnational science in the Arab world

The innovative research described by this book stands in contrast to many bleak portrayals of science in the contemporary Arab world. Many of these portrayals rely on quantitative assessments, such as numbers of scientists, published papers and patents, and gross expenditure on research and development. Based on this expenditure, the Arab region 'performs miserably' with 0.2 per cent of its gross domestic product, wrote Nader Fergany in 2000.[25] The United Nations' *Arab Human Development Report 2003*, with the subtitle 'Building a knowledge society', tells 'a story of stagnation', especially in science. 'In addition to thin production', the report says, 'scientific research in Arab countries is held back by weak basic research and the almost total absence of advanced research in fields such as information technology and molecular biology.'[26] Another study from 2010 calculated that the entire Arab region counted for little more than 1 per cent of the world's production of science. The study also diagnosed 'a clear under-specialisation in life sciences' in Arab countries. Egypt, the largest scientific producer in the Arab world, was said to have 'a deficit in biology', which was 'very deep in fundamental biology'.[27]

While informative, such quantitative assessments of the Arab world's share in global scientific production also have limitations. Quantifications often rely on selective citation bases that exclude many journals published in the Third World.[28] In addition, citation databases favour professional scientists and institutions over other actors in the production of science. They thus exclude 'researchers in the wild',[29] such as Bedouin collaborators and informants as well as non-human actors, such as plants, animals and machines, which do not appear among the lists of authors. Citation counts also tend to focus on an author's official affiliation, even if she undertook field research elsewhere. This was the case in papers authored by Joe Williams, whose official affiliation was Ohio State University. Even when he acknowledged that a paper reported research undertaken at the NWRC, this paper likely counted as an American rather than a Saudi publication. Arguably, evaluations of 'Arab human development' should pay more consideration to the transnational nature of much of modern and contemporary science. Finally, citation databases per definition focus on a very narrow definition of 'output' in the form of publications or patents, ignoring the production of other knowledge, especially in oral and unpublished forms.

Another problem with negative assessments of contemporary Arab science has been a focus on research undertaken at public universities. Heavy teaching loads, poor libraries, excessive red tape and limited intellectual stimulation have characterised many Arab public universities.[30] In the state universities of the Gulf, authoritarian and bureaucratic models of governance have also discouraged innovations.[31] However, as this book has demonstrated, much innovative biological research has taken place in islands of efficiency independent of public universities, such as wildlife and agricultural research centres. In these institutions, scientists often enjoyed better funding than at Western research institutes as well as much logistical support.

One might consider the research that took place at scientific islands of efficiency such as the NWRC, not as 'Arab' or 'Gulf' science, as it was often undertaken by expatriates. However, I argue that it has been part of transnational science *in* the Gulf and the Arab world. The transnational movement of scientists has been beneficial to the growth of scientific communities in many countries. The United States, for instance, hired many foreign scientists during the course of the twentieth century. In the 1930s, it absorbed many scientists who had fled Nazi Germany. Similarly, after the proclamation of the People's Republic of China in 1949, about 4,000 Chinese students and scientists decided to stay in the United States, further contributing to the 'transnationalization of the American scientific community'.[32] Some of these Chinese scientists, just like many European-born American scientists, then went on to win Nobel Prizes. While some of them might not have self-identified as 'Americans', their work was part of transnational science *in* America. As Western economic growth, including growth in higher education and science, partly relied on Middle Eastern oil, the Gulf contributed to this transnational science.

Due to limited access to citizenship,[33] very few expatriate scientists in the Gulf became citizens of their host countries as compared with naturalised academics in the United States. However, many foreign biologists worked and lived in the Gulf for years or decades. They just formed part of the Gulf's multinational societies for considerable periods. The Pakistani botanist Shaukat Chaudhary, for instance, worked at the National Agriculture and Water Research Center from 1978 until his retirement in 2001, and remained an honorary taxonomist emeritus at the centre thereafter. At the Gulf's universities, numerous Egyptians and other expatriates taught and researched biology over decades. Some expatriate biologists also resided in more than one Gulf country, thus contributing to an integration of Gulf science. The Egyptian botanist Kamal Batanouny, for instance, worked at King Abdulaziz University in

Saudi Arabia and subsequently at Qatar University. Similarly, the German zoologist Friedhelm Krupp worked in Saudi Arabia over decades, before becoming the founding director of the Qatar Museum of Natural History in 2011.[34]

Other foreign scientists did not become residents of the Gulf monarchies, but still conducted much of their research inside Arabia. Even though they had foreign affiliations, this research was thus still part of transnational science *in* the Gulf. Joe Williams of Ohio State University, for instance, was not formally employed by a Gulf institution, but undertook extensive field research in the kingdom in cooperation with local colleagues, like the Saudi ornithologist Mohammed Shobrak. He perhaps also connected to wider Gulf society in a more direct and riskier way than many of the expatriates formally employed by scientific islands of efficiency did. Williams lectured at Taif University and, while his lecture caused a controversy, it demonstrated his presence in the Gulf's public institutions and debates. By lecturing in Saudi Arabia and publishing in the *Saudi Journal of Biological Sciences*, Williams participated in the transnationalisation of Gulf science.

Scientific islands of efficiency in the Gulf were not restricted to specialised research centres, but also encompassed a number of universities established in recent years. Among the most prominent is the King Abdullah University of Science and Technology, which opened in 2009. Another island of efficiency in the Gulf's public sector, Saudi Aramco, established KAUST under the patronage of the monarch himself. Its funding was protected through a religious endowment and thus independent from the Ministry of Higher Education. Staffed by highly paid foreign experts and located outside major cities, it enjoyed autonomy from large parts of the Saudi state and society. However, this isolation and KAUST's character as a personalised initiative also makes it potentially vulnerable to changes in the kingdom's leadership.[35] Other scientific islands of efficiency include the branch campuses of foreign universities in the Gulf, including those located in Qatar's Education City. They also enjoy patronage from key members of the ruling families, managerial autonomy, and the funding to hire highly paid foreign experts. However, their future often also depends on the continuation of patronage and funding by institutions such as Qatar Foundation.

Transnational science in the Gulf was, however, not just restricted to institutional islands of efficiency. If 'nature' itself was the 'laboratory' for

Darwinian plant ecology in the German Empire,[36] then Gulf environments also formed an important site of colonial and postcolonial ecological studies. Arabia provided an important field for global biological research, even if some of the scientists returned to universities, museums or botanical gardens in Europe and America after their work. As a transit point between Africa and Asia, the Arabian peninsula held important fossils for evolutionary studies on humans and other animals. As a site of high temperatures and scarcity of water, Arabia also provided a venue for the study of adaptations to extreme environments and climate change. If we consider flora, fauna and the climate as nonhuman actors, the environment of the Gulf itself has made major contributions to global science.

## The future of Gulf biology

Given the long history of transnationalism in the Gulf,[37] Gulf science probably continues to be transnational. Some of the future connections will likely be with scientists in Western as well as Arab countries. In 2011, the Mammals Research Chair at King Saud University signed a cooperation agreement with the Zoological Society of London[38] that promised to extend the long history of cooperation between the society and Saudi wildlife institutions into the future. In addition to long-term partners in Britain or Egypt, Morocco may come to play a more important role. Sharing a monarchical political system and Arab and Muslim cultural aspects, Morocco had long hosted princes from the Gulf for extended stays. Threatened by revolution, Morocco and the countries of the Gulf Cooperation Council further strengthened their political ties across the geographical distance during the course of the Arab Spring or uprisings. In 2011, the Gulf monarchies invited Morocco, together with Jordan, to apply for membership in the GCC.[39]

Scientific connections will likely add to the political ties. Morocco had a climate comparable to that of the Gulf, which allowed for similar research initiatives in the environmental sciences. One initiative has been the Emirates Center for Wildlife Propagation (ECWP), which was founded in eastern Morocco in 1995. Its original sponsor was Zayed bin Sultan, who had previously established the National Avian Research Center in Abu Dhabi. After Sheikh Zayed's death in 2004, his son Mohammed's International Fund for Houbara Conservation continued to support the two centres in Morocco and Abu Dhabi. Like NARC, the ECWP aimed at producing self-sustaining houbara populations.[40]

The Emirates Center for Wildlife Propagation formed part of a growing transnational network of rentier science that focused on houbara breeding. Jacques Renaud's company Reneco, which had previously managed the NWRC near Ta'if, ran the ECWP. Like Renaud, a number of scientists who had previously worked for the NWRC staffed the ECWP. With the rich experience of these experts,[41] the Moroccan centre achieved successes very quickly, producing 151 chicks in 1999 and almost 800 in 2002. A protection programme accompanied this breeding. In 1997, the ECWP created a protected area of a thousand square kilometres, which it extended by 14,000 square kilometres in 2002. In the following year, Reneco started managing further outposts of rentier science in Central Asia, which formed an important habitat for houbara bustards. Between 2009 and 2012, the company became responsible for a Sheikh Khalifa Houbara Breeding Center in Kazakhstan and an Emirates Center for Conservation of Houbara in Uzbekistan.[42]

Located outside the Gulf entirely, the researchers at the ECWP perhaps enjoyed even more freedom to engage in sensitive research than their colleagues in scientific islands of efficiency in Abu Dhabi did. At the same time, the ECWP's location in Morocco allowed for research that might raise ethical questions about experiments harmful to animals if conducted in Europe. Besides breeding birds, the ECWP investigated the breeding behaviour and evolutionary strategies of houbara bustards in collaboration with the Muséum national d'Histoire naturelle in Paris. In a study of sexual selection, the researchers deliberately infected male houbaras with a bacterium and found that these houbaras had a poorer courtship display. Females who mated with these males were at risk of contracting infectious diseases or not producing viable offspring. Based on these observations, the scientists concluded that female choice reinforced the 'evolution of male sexual display'.[43]

Apart from connections with North Africa and Central Asia, East Asian connections will likely become more important for Gulf biology. The spectacular growth of the East Asian economies, in particular of China, from the 1980s onwards – just like the reconstruction and prosperity of Western Europe after World War II – relied on Middle

Eastern oil. Gulf rulers also increasingly realised the political importance of East Asian countries. The Persian Gulf and Pacific Asia thus moved from a previous phase of 'indifference' to 'interdependence'.[44] Already in 1990, Prince Saud Al-Faisal, the Saudi foreign minister and managing director of the NCWCD, visited China. In 1998, the Saudi Crown Prince Abdullah followed him.[45] As Chinese and other East Asian countries are increasing scientific as well as economic production, a strengthening of scientific connections will likely follow the strengthening of political links in the future. Reflecting the recognition of Asian scientific institutions in the Gulf and globally, KAUST's first president was a former president of the National University of Singapore and chair of the Association of Pacific Rim Universities.[46]

Gulf institutions found partnerships with China particularly attractive in the field of genomics. In the 2000s, the King Abdulaziz City for Science and Technology in Riyadh started collaborating with the Beijing Genomics Institute (BGI). This was part of 'KACST's strategy trying to link itself to different organisations and institutes in different countries based on their achievements', explained one Saudi scientist.[47] One of the first major initiatives of this partnership was to sequence the camel's genome. Besides its symbolic value, Saudi scientists chose the species for its agricultural potential. KACST was especially interested in improving the breeding of camels and the quality of their meat and milk.[48] However, like other biologists in the Gulf, KACST was also interested in animal adaption to the harsh desert environment. 'Unlocking the camel genome will facilitate the identification of genes that explain how camels adapt to their habitat', said another Saudi researcher involved.[49] In 2010, KACST and BGI announced that they had sequenced all 2.2 billion nucleotides in the Arabian camel's genome, revealing a strong similarity between camels and cattle.[50]

The sequencing of the camel's genome preceded that of another symbolically and agriculturally important organism, the date palm. Saudi Arabia alone not only displayed the date palm in its coat of arms, but also cultivated more than 25 million palm trees and produced 15 per cent of the world's date production. In sequencing the genome, Gulf institutions and their international collaborators even competed with each other. In 2008, the King Abdulaziz City for Science and Technology started a four-year Date Palm Genome Project, again in collaboration with the Beijing Genomics Institute.[51] At about the same time, the Weill Cornell Medical College in Qatar, a country that also had the date palm in its coat of arms, undertook research on the date palm's genome too. Creating another transnational initiative, Cornell collaborated with researchers from the Qatari Ministry of Environment, universities in America and the French Centre national de la recherche scientifique. In parallel, the KACST- and Cornell-led teams published articles on different parts of the date palm's genome.[52]

In their genomic research, the scientists also referred to evolution. In 2012, the KACST and BGI scientists published an article on the mitochondrial genome of the date palm. In their introduction, the authors explained that sequencing the mitochondrial genome was essential for 'improving its agricultural, horticultural, and nutritional values'. The scientists also claimed that their analysis was 'of importance in revealing mechanisms underlying mitochondrial genome evolution and the unique evolutionary status' of the date palm among flowering plants. The scientists found the date palm particularly interesting, as the mitochondrial genomes of the other flowering plants were 'known for their slow evolutionary rate'.[53]

So far, much of the recent research in genomics and evolutionary biology has remained rather uncontroversial. However, it is likely that at least some controversies will arise in the future. The German zoologist and editor of the *Fauna of Saudi Arabia* series Friedhelm Krupp even expected them to do so. He stated that he wanted to include evolution in the exhibits of the Qatar Natural History Museum, of which he was the founding director. Among other topics, he planned exhibits on dinosaurs and the evolution of the Arabian flora and fauna. Moreover, he planned to show the cabinet of the nineteenth-century British naturalist Alfred Russell Wallace. As Wallace had 'discovered the theory of evolution in parallel to Darwin', Krupp 'expected problematic discussions'.[54] In Saudi Arabia, researchers are also expecting to discover more traces of human ancestors who passed through a more fertile Arabia in prehistoric times while moving between Africa and Asia. In 2014, Prince Sultan bin Salman, the president of the Saudi Commission for Tourism and Antiquities, even pledged to create a Green Arabia Institute devoted to research on the peninsula's early occupation.[55] This institute could well form another scientific island of efficiency under princely patronage.

Whether or not the theory of evolution will become more controversial, this book has shown that modern biology has established itself in the Gulf monarchies. This field of science did not rely on local professional biologists alone, but also on connections with foreign scientists and institutions, 'researchers in the wild' and nonhuman actors. These networks made the Gulf monarchies part of transnational and global science, rather than a peripheral region. At the same time, the economic and social dynamics of the Gulf monarchies produced a form of rentier science characterised by patron–client relations and islands of efficiency, autonomy and academic freedom amidst more controlled and censored institutions. How long rentier science will last is as difficult to predict as the longevity of the oil-and gas-driven economies. However, even if fossil fuels are exhausted, the Arabian peninsula will likely continue to hold important resources for evolutionary research in the form of living and extinct creatures. These resources will probably attract scientists to the region, whatever the fate of the rentier states.

# NOTES

## Acknowledgements

1. Angus Stevenson, *Oxford Dictionary* (2010).
2. Library of Congress, 'ALA-LC Romanization' (2010).
3. University of Chicago Press, *The Chicago Manual* (2010), 660–784.

## Chapter 1    Scientific Gulf

1. Gulf News, 'Saudis' (2001). Other reasons for banning Pokémon included 'hidden symbols of Zionism'. Guessoum, *Islam's Quantum Question* (2011), 273.
2. BBC, 'Qatari religious leader' (2001).
3. Skovgaard-Petersen and Gräf, *The Global Mufti* (2009).
4. BBC, 'Qatari religious leader' (2001).
5. Smalley, 'Dubai scholars' (2001).
6. BBC, 'Qatari religious leader' (2001).
7. Al-Lajnah al-Dā'imah lil-Buḥūth al-'Ilmīyah wa-al-Iftā', 'Al-taṭawwur' (2014).
8. Ibid., 'Naẓarīyat al-taṭawwur' (2014).
9. Al-Qaraḍāwī and 'Uthmān, 'Bidāyat al-khalq' (2011).
10. Quoted in: Burton, 'Teaching' (2010), 26.
11. Qatari student, personal communication, 7 September 2014.
12. Krupp, interview, 14 January 2013. Abu Bakr Bagader, interview, 14 June 2013.
13. Shrek, 'Ṭālibāt' (2007).
14. Abū Zinādah, 'Liqā'' (1973).
15. Abuzinada, 'The Role' (2003), 39; 'Restoration' (1998), 641.
16. Ewers and Malecki. 'Leapfrogging' (2010).
17. Kruk, 'Ibn Abī l-Ash'ath's *Kitāb al-Ḥayawān*' (2001). Eisenstein, 'Some Accounts' (1995).

18. Eisenstein, 'Die arabische Tierkunde' (2007), 154–5.
19. Ibid., 'Die Trappe' (1983).
20. Klaver, *Scientific Expeditions* (2009).
21. Onley, *The Arabian Frontier* (2007).
22. Hayhurst, 'A Quest' (2014).
23. Carter, 'Some Plants' (1917). V. Dickson, 'Plants' (1938).
24. Latour, *Science* (1987), 215.
25. Said, *Orientalism* (1978), 86. Klaver, *Scientific Expeditions* (2009), 42.
26. Michaelis, *Fragen* (1762). I thank Rachel Cohn for providing me with a copy of Michaelis's work.
27. Sörlin, 'National and International Aspects' (1993).
28. Klaver, *Scientific Expeditions* (2009), 158.
29. Parker, 'Controlling' (2012).
30. Toby Jones, *Desert Kingdom* (2010).
31. De Marco and Dinelli, 'First contribution' (1974).
32. McDonald, 'Animal-Books' (1988), 3.
33. V. Dickson, 'Plants' (1938).
34. Mandaville, *Bedouin Ethnobotany* (2011), xi; *Flora* (1990).
35. Waardenburg, *Les universités* (1966), 299–300.
36. Willoughby, *Let a Thousand Models* (2008), 5–6.
37. Saudi Arabian Cultural Mission to the U.S.A., *Directory* (2006), 1–22.
38. Marwa Elshakry, *Reading Darwin* (2013).
39. They include 'Abd al-Quddoos al-Ansari and Muqbil al-Dhukayr. Aramco World. 'Portrait' (1964), 4. Āl Bassām, *Khizānat al-tawārīkh* [1999], 17.
40. Helmy Mohammad, 'Notes' (2000), 251.
41. Ochsenwald, 'Saudi Arabia' (1981), 272.
42. Bāshumayyil, *Al-Islām* (1964). Al-Omar, 'The Reception' (1982), 343–4.
43. Stéphane Lacroix, *Les islamistes* (2010). Atar, 'Quest for Identity' (1988).
44. Atwan, 'In the realm' (1996), 79.
45. Ipsos, 'Ipsos' (2011).
46. On relativity, see Ziadat, 'Early Reception' (1994).
47. Abdullah Al-Omar, 'The Reception' (1982). Ayoub, 'Creation' (2005). Helmy Mohammad, 'Notes' (2000). Howard, *Being Human* (2011). Marwa Elshakry, *Reading Darwin* (2013); 'Early Arabic Views' (2012); 'Muslim Hermeneutics' (2011); 'Darwinian Conversions' (2010); 'The Gospel' (2007). Riexinger, 'Responses' (2009); 'Islamic Opposition' (2010). Ziadat, *Western Science* (1986).
48. Asghar et al., 'The origin' (2010). Burton, 'Teaching' (2010); 'Evolution' (2010), 'Science' (2010). BouJaoude et al., 'Biology Professors' and Teachers' Positions' (2011); 'Muslim Egyptian and Lebanese Students' Conceptions' (2011). Dagher and BouJaoude. 'Scientific Views (1997). Peker et al., 'Three Decades' (2010). Vlaardingerbroek and Hachem-El-Masri. 'The Status' (2006).
49. Edis, *An Illusion* (2007). Marwa Elshakry, 'The Exegesis' (2008). Guessoum, *Islam's Quantum Question* (2011). John Livingston, 'Western Science' (1996). Stenberg, *The Islamization* (1996). Stolz, 'By virtue' (2012).

50. Toby Jones, *Desert Kingdom* (2010). Mitchell, *Colonising* (1988), *Rule* (2002). Schayegh, *Who Is Knowledgeable* (2009). Omnia El Shakry, *The Great Social Laboratory* (2007). Beaudevin, 'Old diseases' (2013).
51. E.g., Lotfalian, *Islam* (2004).
52. Dallal, *Islam* (2010), 158–9, 175. Bond, 'Where progress' (2003).
53. Aydin, 'Introductory Essay' (2010).
54. See, e.g., Badran, 'The Arab States' (2005), 161; King, *Islamic Astronomy* (2012), xii; Maziak, 'Science' (2005), 1417; Sezgin, *Wissenschaft* (2003), 168.
55. Abdullah Al Saud, 'King's Speech' (2009).
56. Ahuja, 'A new golden age' (2012).
57. Toby Jones, *Desert Kingdom* (2010), 241–2.
58. Davidson, *After the Sheikhs* (2012), 104.
59. Ibid.
60. Lewin, 'U.S. Universities' (2008).
61. Butler, 'Can a Saudi university' (2007), 759.
62. Sardar, 'Middle East brain drain' (1980).
63. Edis, 'Modern Science' (2009).
64. Al-Atawneh, *Wahhābī Islam* (2010), xvi.
65. Lacey, *Inside the Kingdom* (2009), 89–90. Al-Azm, 'Islam' (2007), 292. Ende, 'Religion' (1982), 381.
66. Dobzhansky, 'Nothing' (1973), 125.
67. Matthiesen, *Sectarian Gulf* (2013).
68. Golshani, 'Islam' (2000).
69. Bucaille, *La Bible* (1976).
70. Marwa Elshakry, 'Global Darwin' (2009), 1200.
71. On this notion, see Marwa Elshakry, 'When Science' (2010).
72. Stenberg, *The Islamization* (1996), 13.
73. Faheem, 'Higher Education' (1982), 199, 41, 188.
74. Nasseef, 'Role' (1992), 3–4.
75. Al-Atawneh, *Wahhābī Islam* (2010), 59.
76. Guessoum, 'The Qur'an' (2008), 420–3.
77. Government of Dubai Media Office, 'Dubai International Holy Quran Award' (2013).
78. Butler, 'Can a Saudi university' (2007), 758.
79. Mazawi, 'Divisions' (2003), 248–9.
80. Tétreault, 'Identity' (2011), 93.
81. Atwan, 'In the realm' (1996), 79.
82. Faheem, 'Higher Education' (1982), 139.
83. Nazim Ali, 'Acquisition' (1989), 110.
84. Teitelbaum, 'Dueling' (2002), 224.
85. Al-Atawneh, *Wahhābī Islam* (2010), 120.
86. Nazim Ali, 'Acquisition' (1989), 110.
87. Torstrick and Faier, *Culture* (2009), 53–4.
88. Herb, *All in the Family* (1999).

89. Commins, *The Gulf States* (2012), 120. Anscombe, *The Ottoman Gulf* (1993).
90. Fromherz, *Qatar* (2012).
91. Beblawi and Luciani, *The Rentier State* (1987).
92. Hertog, 'The Sociology' (2010), 313. On the Gulf monarchies as rentier states, see also Springborg, 'GCC Countries' (2013).
93. For an overview, see Gray, *A Theory* (2011). See also Mitchell, *Carbon Democracy* (2013).
94. Beblawi, 'The Rentier State' (1987), 385–6.
95. United Nations Development Programme, *The Arab Human Development Report* (2003), 135.
96. Bizri, 'Research' (2013), 196.
97. Al-Misnad, *The Development* (1985).
98. Al-Toraif, *Convention* (2009), 9.
99. Marwa Elshakry, 'Global Darwin' (2009).
100. Basalla, 'The Spread' (1967).
101. Fan, 'The Global Turn' (2012). Raj, 'Beyond Postcolonialism' (2003); *Relocating* (2007). Lissa Roberts, 'Situating Science' (2009).
102. Edis, 'Modern Science' (2009), 886–8.
103. Marwa Elshakry, 'The Gospel' (2007), 207–9.
104. Edis, *An Illusion* (2007), 128; 'Cloning' (1999). Hameed, 'Evolution' (2001), 141. Riexinger, 'Propagating' (2008), 103. Elise Burton, personal communication, 10 March 2013.
105. Milgram, 'The Small-World Problem' (1967). Granovetter, 'The Strength' (1973). Recent overviews of research on social networks include Kadushin, *Understanding* (2011); Degenne and Forsé, *Introducing* (1999).
106. Latour, *Science* (1987); *Pandora's Hope* (1999).
107. Barabási et al., 'Evolution' (2002). Newman, 'The structure' (2001). Wagner and Leydesdorff, 'Network structure' (2005); 'Mapping' (2005). Leydesdorff and Wagner, 'International Collaboration' (2008).
108. Beblawi and Luciani, *The Rentier State* (1987). Gray, *A Theory* (2011).
109. Hertog, *Princes* (2010); 'The Sociology' (2010).
110. Ibid., 'Shaping' (2007), 548.
111. McNeill and McNeill, *The Human Web* (2003).
112. Lester, 'Imperial Circuits' (2006). Glaisyer, 'Networking' (2004).
113. Bennett and Hodge, *Science* (2011). Ballantyne, 'Race' (2001). Harris, 'Long-Distance Corporations' (1998). Lux and Cook, 'Closed Circles' (1998).
114. See, for instance, Andersen et al., 'The Money Trail' (2012).
115. The notion of the Middle East as a scientific periphery also appears in sociological studies, like Wagner and Leydesdorff, 'Mapping' (2005).
116. Potter, 'Webs' (2007), 646.
117. Findlow, 'International Networking', (2005), 290.
118. Ibid., 291.
119. Ninham, *A Cohort* (2012).
120. Mazawi, 'Divisions' (2003), 107.

121. Drayton, *Nature's Government* (2000).
122. Latour, *Science* (1987).
123. David Dickson, 'Princeton' (1980).
124. Breakspear, 'Refereeing' (1984).
125. Al-Kobaisi, 'Joint research project' (1983).
126. Ménoret, *Joyriding* (2014), 9.
127. Sabra, 'Situating' (1996).
128. Temple, Letter (1950). Gallagher, Letter (1977, 1979).
129. Drayton, *Nature's Government* (2002), 266.
130. Desmond, *The History* (2007), 287.
131. Akasoy, 'The Influence' (2011), 410.
132. Davidson, *After the Sheikhs* (2012), 75.
133. Grutz, 'A King' (2008), 45.
134. Burchard, Letter (1980; 1981).
135. Basson et al., *Biotopes* (1977), 19.
136. Qatar Digital Library, 'About' (2014).
137. Kohler, 'Place' (2002), 191.
138. Abū Zinādah, *Al-masīratān* (2011). Whybrow, 'Brains' (2000).
139. Scott, *Seeing* (1998).
140. Ibid.
141. Darwin, *On the Origin* (1859), 20–43.

## Chapter 2    Plant Kingdoms

1. Chaudhary and Al-Jowaid, *Vegetation* (1999), 17.
2. Ouis, 'Greening' (2002).
3. Davidson, *Abu Dhabi* (2009), 137.
4. Toby Jones, *Desert Kingdom* (2010), 5.
5. Prokop, 'Political Economy' (1999), 117.
6. Toby Jones, *Desert Kingdom* (2010), 63–85.
7. Wazārat al-Maʿārif, *Ahamm* (2000), 109.
8. Bismarck, *British Policy* (2013).
9. Hodge, 'British Colonial Expertise' (2010); 'Science' (2002).
10. Chaudhary, email, 31 January 2013.
11. Jones, *Desert Kingdom* (2010).
12. Prokop, 'Political Economy' (1999), 163.
13. Dundar, 'Empire' (2014).
14. Marwa Elshakry, *Reading* (2003). Andrew Jones, *Developmental Fairy Tales* (2011).
15. Chaudhary and Al-Jowaid, *Vegetation* (1999), [13].
16. Ibid., 30.
17. Quoted in: Burton, 'Teaching' (2010), 26.
18. Toby Jones, *Desert Kingdom* (2010).
19. Buj, 'International experimentation' (1995).

20. Mandaville, *Flora* (1990), 4.
21. Vitalis and Heydemann, 'War' (2000), 116–20.
22. Khattab and El-Hadidi, *Results* (1971), 3–4.
23. Ibid.
24. Roman, *Emigration Policy* (2006), 1.
25. Abū Zinādah, *Al-masīratān* (2011), 34. Dan Stolz kindly pointed out to me that the term *ustādh al-jīl* in Egypt refers to the intellectual and first director of Cairo University Ahmad Lutfi al-Sayyid (1872–1963). However, the epithet could be attached to others as well.
26. Migahid, *Migahid and Hammouda's Flora* (1978). Migahid et al., *Proceedings* (1977). Saudi Biological Society, *Proceedings* (1978).
27. Batanouny, *Ecology* (1981), xiii.
28. Ibid., *Plants* (2001), 1.
29. Al-Batānūnī, *Nabātāt* (1986), [219].
30. Batanouny, *Natural History* (1978), xii.
31. Tamraz, 'A Study' (1984), 21.
32. Batanouny, *Natural History* (1978), xii.
33. Ibid., *Ecology* (1981), xii–xiv.
34. Ibid., *Ecology* (1981), xiii.
35. Ibid., *Ecology* (1981), xii.
36. Abdel Bari, 'The Flora' (2005), 114.
37. Batanouny, *Ecology* (1981), xii.
38. Cope, interview, 20 September 2012.
39. Abdel Bari, 'The Flora' (2005), 114.
40. Zahlan, *Science* (1980), 59–60.
41. Boulos, *The Weed Flora* (1988), v.
42. Cope, interview, 20 September 2012.
43. Abdel Bari, 'The Flora' (2005), 114.
44. Kuwait University, 'KTUH History' (2009). Cope, interview, 20 September 2012. See also National Scientific and Technical Information Center, *Iraqi aggression* (1999).
45. Cope, *A Key* (1985).
46. Miller and Cope, *Flora* (1996), ix.
47. Ibid., 25, 359.
48. Cope, interview, 20 September 2012.
49. Kritsky, 'Teaching Evolution' (1984).
50. Boulos, *The Weed Flora* (1988), x.
51. Harbinson, 'The US–Saudi Arabian Joint Commission' (1990), 269.
52. Lippman, 'Cooperation' (2009).
53. Ibid.
54. Dolbee and Gratien, 'Nation' (2012).
55. Comptroller General of the United States, *The U.S.–Saudi Arabian Joint Commission* (1979), 27–8.
56. Chaudhary, interview, 31 January 2013.

57. Collenette, *Wildflowers* (1999), xxxii.
58. Chaudhary, *Grasses* (1989), 10.
59. Collenette, *Wildflowers* (1999), xxx.
60. Al-Rasheed, *A Most Masculine State* (2013).
61. Livingstone, *Putting Science*, 45.
62. Le Renard, 'Only for Women' (2008).
63. Collenette, interviews, 20 and 27 January 2013.
64. Saghir and Chaudhary, *Weed Control Handbook* (1985), 1, 6.
65. Chaudhary and Zawawi, *A Manual* (1983).
66. Saghir and Chaudhary, *Weed Control Handbook* (1985), 1–2.
67. Chaudhary and Akram, *Weeds* (1987).
68. Ibid., vii.
69. Saghir and Chaudhary, *Weed Control Handbook* (1985), 1.
70. Ibid.
71. Collenette, interview, 27 January 2013.
72. Chaudhary, interview, 31 January 2013.
73. Ibid.
74. Chaudhary, *Flora* (1999).
75. See e.g., Stremmel, 'An Imperial German Battle' (2014).
76. Timler and Zepernick, 'German Colonial Botany' (1987).
77. Drayton, *Nature's Government* (2000), 254.
78. Cittadino, *Nature* (1990).
79. Timler and Zepernick, 'German Colonial Botany' (1987).
80. Hiepko, 'Die Sammlungen' (1990).
81. Timler and Zepernick, 'German Colonial Botany' (1987), 158–9.
82. Orth, 'Das Förderprofil' (2004), 273.
83. Gießner, 'Das Afrika-Kartenwerk' (1969).
84. Röllig, 'Der Tübinger Atlas' (1974), 531.
85. E.g., Zohary, 'On the Flora' (1981).
86. Frey and Kürschner, *Tübingen Atlas* (1985), 10, 19.
87. Ibid. and Probst, *Vegetation* (1978), 10.
88. Ibid., 'A tribute' (2010), 9.
89. Ibid. and Kürschner, *Tübingen Atlas* (1985), 26.
90. Kürschner, interview, 10 October 2012.
91. Frey, interview, 24 October 2012.
92. Ibid. and Kürschner, *Tübingen Atlas* (1985), 9, 24.
93. El-Sheikh, 'International Cooperation' (2012).
94. Frey, interview, 24 October 2012.
95. Kürschner, interview, 10 October 2012.
96. Frey and Kürschner, 'The first records' (1982), 159–60.
97. Fisher et al., *The Natural History* (1999), 1.
98. Frey and Kürschner, *Tübingen Atlas* (1985), 24.
99. Spalton, 'Ralph' (2006), 29.
100. Frey and Kürschner, *Tübingen Atlas* (1985), 24.

101. Ibid., 'Masqaṭ area' (1986), 217.
102. Baierle et al., *Vegetation* (1985).
103. Frey and Kürschner, 'Masqaṭ area' (1986), 215–16.
104. Ibid. and Wolfgang, 'Neue Vorstellungen' (1977), 136, 133.
105. Ibid. and Kürschner, 'The first records' (1982), 158.
106. Ibid., *Tübingen Atlas* (1985), 24, 27–8.
107. Ibid., interview, 24 October 2012.
108. Kürschner, interview, 10 October 2012.
109. Frey, interview, 24 October 2012.
110. Burton, 'Teaching Evolution' (2010).
111. Krupp et al., 'Preface' (1987), 7.
112. Gause, 'Saudi Arabia' (2000), 82.
113. Kürschner, interview, 10 October 2012.
114. Ibid.
115. El-Shibiny, 'Higher Education' (1997), 168.
116. Ghazanfar, interview, 27 September 2012.
117. El-Shibiny, 'Higher Education' (1997), 170–1.
118. Ghazanfar, interview, 27 September 2012.
119. Mazawi, 'Divisions' (2003), 94.
120. Ghazanfar, interview, 27 September 2012.
121. El-Shibiny, 'Higher Education' (1997), 163.
122. Delany, interview, 29 January 2013.
123. Fisher, interview, 17 January 2013.
124. Delany, interview, 29 January 2013.
125. Fisher, interview, 17 January 2013
126. Delany, interview, 29 January 2013.
127. Ibid.
128. Ghazanfar, interview, 27 September 2012. M. B. V. Roberts, *Biology* (1986).
129. Delany, interview, 29 January 2013. Ghazanfar, interview, 27 September 2012.
130. Ghazanfar, interview, 27 September 2012.
131. Fisher, interview, 17 January 2013
132. El-Shibiny, 'Higher Education' (1997), 163.
133. Fisher, interview, 17 January 2013
134. Ghazanfar, 'CV' (2012).
135. Ibid., interview, 27 September 2012.
136. Cope, interview, 20 September 2012.
137. Ghazanfar, 'CV' (2012), 1–2.
138. Ibid., interview, 27 September 2012.
139. Kürschner and Ghazanfar, 'Bryophytes' (1998), 100, 124.
140. Frey and Kürschner, 'Wüstenmoose' (1998).
141. Turkish Journal of Botany, 'Scope' (2013).
142. Kürschner, 'Biogeography' (2008), 433.
143. Ibid., 445.
144. Gaube, interview, 2 February 2013.

145. Lang, 'Heinz Gaube' (2011).
146. Buerkert, 'The Project' (2012).
147. Ibid., interview, 7 January 2013.
148. These views are also reflected in Ghabra and Arnold, *Studying* (2007), 15. Badran, 'The Arab States' (2005), 161.
149. Buerkert, interview, 7 January 2013.
150. Mazawi, 'The Academic Workplace' (2003), 242.
151. Patzelt, interview, 15 January 2013.
152. Ibid.
153. Buerkert, interview, 7 January 2013.
154. Scott, *Seeing* (1998).
155. Buerkert, interview, 7 January 2013.
156. Ibid.
157. Beblawi, 'The Rentier State' (1987), 385.
158. Buerkert, interview, 7 January 2013.
159. Ibid. and Gebauer, *Agrobiodiversity* (2009), 7.
160. Ibid., interview, 7 January 2013.
161. Al Khanjari, *Exploration* (2005), 138.
162. Hammer, interview, 9 January 2013. On the cosmopolitanism of Zanzibari Omanis, see Al-Rasheed, 'Transnational Connections' (2005).
163. Al Khanjari et al., 'Molecular diversity' (2007), 1292.
164. Ibid., 'Morphological spike diversity of Omani Wheat' (2008), 1185.
165. Ibid., *Exploration* (2005), 117.
166. Hammer at al., 'Oman' (2009), 552.
167. Gutberlet, 'Oases' (2007), 1.
168. Ibid., 'Jabel' (2007).
169. Buerkert and Schlecht, *Oases* (2010), 5.
170. Gebauer at al., 'Plant Genetic Resources' (2010), 29, 31.
171. Häser et al., 'From Hunter-Gatherer Communities' (2010), 16.
172. Luomi, *The Gulf Monarchies* (2012).

## Chapter 3     Sultans, Consultants and Conservationists

1. Fauna & Flora International, 'Bahrain' (1980), 231.
2. Clark, 'In the Lions' Den' (1985).
3. Ibid.
4. Eisenstein, *Einführung* (1991), 114, 227.
5. Mikhail, *The Animal* (2014), 6.
6. Eisenstein, *Einführung* (1991), 213.
7. Lowe, 'The King's Oryx' (2014).
8. Mikhail, *The Animal* (2014), 1.
9. Eisenstein, *Einführung* (1991), 213.
10. Clark, 'A Capital Zoo' (1990).
11. Llewellyn, 'The Basis' (2003), 213.Gari, 'A History' (2006), 220.

12. Vincent, *Saudi Arabia* (2008), 276.
13. Davis, 'Environmentalism' (2000).
14. Ibid., 'Enclosing Nature' (2013).
15. Barton, 'Environmentalism' (2010).
16. Grove, *Green Imperialism* (1995). On the comparative case of North Africa, see Davis, *Resurrecting* (2007).
17. Bagader et al., *Environmental Protection* (1994), viii.
18. Luomi, *The Gulf Monarchies* (2012).
19. Vincent, *Saudi Arabia* (2008), 254.
20. Foltz, 'Islamic Environmentalism' (2003), 258.
21. Ba Kader et al., *Basic Paper* (1983). Bagader et al., *Environmental Protection* (1994). Bagader, interview, 13 June 2013.
22. Foltz, 'Introduction' (2005).
23. Bagader et al., *Environmental Protection* (1994), 1–3.
24. Frankel and Soulé, *Conservation* (1981), 6, 1.
25. Aramco World, 'A Zoo' (1961).
26. Nyhart, *Modern Nature* (2009), 25–6.
27. Lowe, 'The King's Oryx' (2014).
28. Prendergast and Adams. 'Colonial Wildlife Conservation' (2003): 251.
29. Krawietz, 'Falconry' (2013).
30. Teller, 'Rx' (2009).
31. Time, 'A Kingdom' (1963), 54.
32. Hoye, 'Return' (1982).
33. Abu-Zinada et al., 'The Arabian Oryx Programme' (1988), 41–2.
34. Hoye, 'Return' (1982).
35. Grimwood, '"Operation Oryx"' (1964), 223.
36. Time, 'A Kingdom' (1963), 54.
37. Stanley Price, *Animal Re-introductions* (1989), 47.
38. Grimwood, '"Operation Oryx"' (1964), 223–4.
39. Ibid., 224.
40. E.g., Aramco World, 'A Zoo' (1961). Hoye, 'Return' (1982). Clark, 'In the Lions' Den' (1985).
41. Jungius, 'Plan' (1978), 329–30.
42. Spalton et al., 'Arabian oryx reintroduction' (1999), 169.
43. Percy, 'Obituary' (2006).
44. Daly, '"And What"' (1985), 25.
45. Spalton, 'Ralph' (2006).
46. Hoye, 'Return' (1982).
47. Jungius, 'Plan' (1978).
48. Hoye, 'Return' (1982).
49. Teller, 'Rx' (2009).
50. Daly, '"And What"' (1985), 25.
51. Stanley Price, *Animal Re-introductions* (1989), 59–62.
52. Hoye, 'Return' (1982).

53. Spalton, 'Ralph' (2006).
54. Daly, '"And What"' (1985), 25.
55. Hoye, 'Return' (1982).
56. Daly, '"And What"' (1985), 26.
57. Spalton et al., 'Arabian oryx reintroduction' (1999), 171.
58. UNESCO, 'World Heritage List' (2013).
59. Teller, 'Rx' (2009).
60. Spalton, 'Ralph' (2006).
61. Daly, '"And What"' (1985), 26.
62. Hoye, 'Return' (1982).
63. Stanley Price, *Animal Re-introductions* (1989), 34.
64. Teller, 'Rx' (2009).
65. Spalton, 'Ralph' (2006), 29.
66. Teller, 'Rx' (2009).
67. Spalton, 'Ralph' (2006), 29.
68. Teller, 'Rx' (2009).
69. UNESCO, 'Arabian Oryx Sanctuary' (2013).
70. Teller, 'Rx' (2009).
71. Hoye, 'Return' (1982).
72. Clark, 'In the Lions' Den' (1985).
73. Mitchell, *Rule* (2002), 46–53.
74. Krupp and Mahnert, 'Introduction' (1996), 2.
75. Wittmer and Büttiker, 'Acknowledgements' (1981), 3.
76. Büttiker and Krupp, 'Introduction' (1994).
77. Krupp and Mahnert, 'Introduction' (1996), 2.
78. Büttiker and Krupp, 'Introduction' (1984), 2.
79. Krupp, interview, 14 January 2013.
80. Büttiker, 'Further Notes' (1981), 6.
81. Krupp and Mahnert, 'Introduction' (1996), 2–3.
82. Ibid., 2.
83. Büttiker and Krupp, 'Introduction' (1985), 3.
84. Banaja, 'Introduction' (1981).
85. Bates, 'Prince Charles' (2004).
86. Abu-Zinada et al., 'The Arabian Oryx Programme' (1988), 42.
87. Ady, 'The National Wildlife Research Center' (1997), 1.
88. Renaud, 'Jacques' (2013).
89. Abū Zinādah, *Al-masīratān* (2011), 70.
90. Ibid.
91. According to Friedhelm Krupp, Abdulbar Al-Gain proposed the NCWCD. Comment on a draft of this chapter, September 2013.
92. Collenette, interview, 27 January 2013.
93. Vincent, *Saudi Arabia* (2008), 248.
94. Hertog, 'How the GCC' (2012), 80.
95. Vincent, *Saudi Arabia* (2008), 261.

96. Abū Zinādah, *Al-masīratān* (2011), 73.
97. Joma, 'The Earth' (1991), 376.
98. Hertog, *Princes* (2010), 13.
99. Joma, 'The Earth' (1991), 376–7.
100. Al-Faisal, 'Preface' (1988).
101. Bin Faisal, 'Preface' (1993).
102. Abuzinada, 'The role' (2003), 40, 42.
103. Gari, 'A History' (2006), 225.
104. Abuzinada, 'The role' (2003), 44.
105. Flamand, 'Arabian Oryx' (1994).
106. Hertog, 'Defying' (2010), 267.
107. Abū Zinādah, *Al-masīratān* (2011), 77.
108. Greth and Williamson, 'Recent Developments' (1996), 2.
109. Asmodé, 'Taïf' (1986), 21.
110. Joseph Williams, email, 16 November 2012.
111. Greth and Williamson, 'Recent Developments' (1996), 2.
112. Another way was giving permits for archaeological excavations to teams from different countries.
113. ZSL, 'Station Manager' (1988); 'Project Manager' (1988).
114. Asmodé and Khoja, 'Arabian Oryx Captive Breeding' (1989), 2.
115. Greth et al., 'An outbreak' (1994), 165.
116. Thouless, 'Conservation' (1991), 224.
117. Greth et al., 'An outbreak' (1994), 167.
118. Abuzinada, 'Captive Breeding' (1995), 22.
119. Büttiker and Krupp, 'Introduction' (1989), 2.
120. Al-'Ubaydān, 'Khādim' (1998).
121. Thouless, 'Conservation' (1991), 227–8.
122. Al-Gain, 'Preface' (1985).
123. Sultan bin Abdulaziz Al Saud, 'Preface' (1994).
124. Ibid.
125. Krupp et al., 'Introduction' (2006), 4.
126. Büttiker and Krupp, 'Introduction' (1985), 2.
127. Talhouk, 'Introduction' (1980), 2.
128. Klausewitz, 'Evolutionary History' (1989), 333.
129. Ibid.
130. Krupp, interview, 14 January 2013.
131. Thouless, 'Conservation' (1991), 227.
132. Harrison and Bates, 'On the occurrence' (1984).
133. Bates, 'Desert Specialists' (1995).
134. Flamand, 'Arabian Oryx' (1994).
135. Viré, 'Ķird' (1986), 131.
136. Teller, 'The happy ones?' (2012).
137. Kruk, 'Traditional Islamic Views' (1995), 30–1.
138. Viré, 'Ķird' (1986), 132.

139. Teller, 'The happy ones?' (2012).
140. Cook, 'Ibn Qutayba' (1999), 46.
141. Teller, 'The happy ones?' (2012).
142. Klaver, *Scientific Expeditions* (2009), 186–7, 193.
143. Kummer, *In Quest* (1995), 233.
144. Ibid., 312.
145. Ibid., 313.
146. Ibid., 313, 317.
147. Ibid., 321–2.
148. Ibid., 322.
149. Arabian Wildlife, 'In Quest' (1997).
150. Biquand et al., 'The Distribution' (1992); 'Group Composition' (1992).
151. Ibid., 'Management' (1994).
152. Gautier and Biquand. 'Primate Commensalism' (1994), 210–1.
153. Klaver, *Scientific Expeditions* (2009), 177, 180.
154. Greth and Williamson, 'Recent Developments' (1996), 2.
155. Ibid. et al., 'The Controversial Subject' (1996), 54–5.
156. Williamson, 'The Relationship' (1996), 49.
157. Ibid., 49–50.
158. Greth and Magin, 'Final Discussion' (1996), 159–60.
159. Ibid. and Williamson, 'Recent Developments' (1996), 3–6.
160. Hammond et al., 'Phylogenetic Reanalysis' (2001).
161. Kummer, *In Quest* (1995), 17.
162. Winney et al., 'Crossing' (2004), 2824.
163. Teller, 'The happy ones?' (2012).
164. Hammond, 'Phylogenetic Reanalysis' (2001), 1131.
165. Bärmann et al., 'The curious case' (2013).

## Chapter 4    Scientific Islands of Efficiency

1. Hellyer and Aspinall, 'Researching' (2005), 18.
2. IFHC, 'The Future' (2011).
3. Aspinall, 'Environmental Development' (2001), 295.
4. Davidson, *After the Sheikhs* (2012), 75–6.
5. Aspinall, 'Environmental Development' (2001), 290.
6. Ouis, 'Greening' (2002), 340.
7. Todorova, '1997' (2011).
8. WWF, 'Emirates Wildlife Society' (2013).
9. Davidson, *Abu Dhabi* (2009), 138.
10. Ibid., 137.
11. Todorova, '1997' (2011).
12. IFHC, 'The Future' (2011).
13. UNESCO, 'Falconry' (2012).
14. IFHC, 'The Future' (2011).

15. On efficient state-owned enterprises, see, Hertog, 'Lean' (2011), 'Defying' (2010), 'How the GCC' (2012).

16. Hertog, 'Lean' (2011), 18.

17. Schulz, interview, 9 January 2013.

18. Ibid.

19. Ibid.

20. Ibid.

21. Ibid.

22. Schwede, interview, 21 January 2013.

23. WWF, 'WWF International Directors' (2013).

24. Scientists who were hired at around the same time included Michel Saint-Jalme, Stephen and Anna Newton, and Olivier Combreau. Yolanda van Heezik, comment on a draft of this chapter, July 2013.

25. Seddon, email, 17 December 2012.

26. Ibid., interview, 27 December 2012.

27. Ibid., interview, 27 December 2012.

28. Seddon et al., 'Restoration' (1995), 139.

29. Islam and Boug, 'Save' (2011), 21.

30. Seddon, interview, 27 December 2012.

31. Ibid., email, 23 June 2013.

32. Larivière and Seddon, 'Vulpes' (2001), 2.

33. Heezik and Seddon, 'Born' (2001), 59.

34. Seddon, interview, 27 December 2012.

35. Satia, 'A Rebellion' (2011), 25–6.

36. Seddon, email, 23 June 2013; interview, 27 December 2012.

37. Ibid., interview, 27 December 2012; email, 23 June 2013.

38. For a comparative case of indigenous and expert knowledge in the Maghrib, see Davis, 'Indigenous knowledge' (2005).

39. Van Heezik, interview, 27 December 2012.

40. Ibid. Joe Williams claimed that van Heezik was concerned with the dangers of driving in Saudi Arabia, adding that Saudis 'typically take more risks while driving'. Comment on a draft of this chapter, June 2013.

41. Beblawi, 'The Rentier State' (1987), 391.

42. Seddon, email, 17 December 2012.

43. Hertog, 'Lean' (2011).

44. Schwede, interview, 21 January 2013.

45. Hellyer, 'The Emirates Natural History Group' (1994).

46. Launay and Paillat. 'A Behavioural Repertoire' (1990). Greth et al., 'Chlamydiosis' (1993).

47. Stéphane Ostrowski, comments on a draft of this chapter, July 2013.

48. Launay, 'Conclusions' (1997), 25.

49. Hellyer, 'In Remembrance' (2011).

50. Hockey and Aspinall, 'The Crab Plover' (1996), 60, 62–3.

51. Ibid. 'Why do Crab Plovers' (1997), 39–40.

52. I was unable to interview Aspinall before he died in 2011.
53. Aspinall and Hockey, 'The Indian Ocean's Crab-loving Plover' (1997).
54. Todorova, 'Emirati' (2013).
55. Davis, 'Brutes' (2008), 258.
56. Ostrowski, email, 15 December 2012.
57. Ibid.
58. Ibid.
59. Ostrowski, email, 15 December 2012.
60. Du Plessis and Williams, 'Communal Cavity Roosting' (1994), 292. Williams, 'A Phylogenetic Perspective' (1996), 469.
61. Ostrowski, email, 15 December 2012.
62. Williams, interview, 3 December 2012.
63. Tieleman, interview, 9 December 2012.
64. Ibid., 'Avian adaptation' (2002), 14.
65. Williams, interview, 3 December 2012.
66. Tieleman, interview, 9 December 2012.
67. David Livingstone, *Putting* (2003), 43.
68. Tieleman, 'Avian adaptation' (2002), 332.
69. Williams, interview, 3 December 2012.
70. Ibid., 'Curriculum Vitae' (2012).
71. Ibid., interview, 3 December 2012.
72. Ibid., interview, 3 December 2012.
73. Tieleman, interview, 9 December 2012.
74. Williams, interview, 3 December 2012.
75. Tieleman, interview, 9 December 2012.
76. Ostrowski, 'Ajustements' (2006).
77. Carey, 'Desert Gazelles' (2006). Garcia, 'Gazelles' (2006).
78. Tieleman, 'Avian adaptation' (2002), 20. See also Williams and Tieleman, 'Physiological Ecology' (2001).
79. Williams and Tieleman, 'Physiological Adaptation' (2005), 416–17.
80. Ibid., 416.
81. Ostrowski, email, 15 December 2012.
82. Williams, interview, 3 December 2012; comment on a draft of this chapter, June 2013.
83. Jennings, *Atlas* (2010), 50.
84. Mazawi, 'Divisions' (2003), 94.
85. Mellahi and Al-Hinai, 'Local Workers' (2000), 177–8.
86. Hertog, *Arab Gulf States* (2014).
87. Friedhelm Krupp, interview, 5 October 2013.
88. Abū Zinādah, *Al-masīratān* (2011), 221.
89. Ibid., 225–6. Joe Williams remembered this episode differently. According to him, Stéphane Ostrowski's wife Katy 'presented a reasonable approach to cutting expenditure'. Yet, Abdul-Rahman Khoja did not 'agree'. Williams, comment on a draft of this chapter, June 2013.

90. Abū Zinādah, *Al-masīratān* (2011), 226–8.
91. Tieleman, interview, 9 December 2012.
92. Ostrowski, email, 15 December 2012.
93. Ibid.
94. Vincent, *Saudi Arabia* (2008), 260.
95. Ostrowski, comments on a draft of this chapter, July 2013.
96. Williams, interview, 3 December 2012. On Al-Faisal's donations to the NCWCD, see also Abū Zinādah, *Al-masīratān* (2011), 222.
97. Beblawi, 'The Rentier State' (1987), 385.
98. Tieleman, interview, 9 December 2012.
99. Mellahi and Al-Hinai, 'Local Workers' (2000), 180.
100. Beblawi, 'The Rentier State' (1987), 388.
101. Williams, interview, 3 December 2012.
102. Shobrak, interview, 19 December 2012.
103. Abū Zinādah, *Al-masīratān* (2011), 284.
104. Ibid., 284–5.
105. Shobrak, interview, 19 December 2012.
106. Ibid.
107. Seddon, email, 23 June 2013.
108. Ibid., interview, 19 December 2012.
109. Williams et al., 'Lizard Burrows' (1999), 717.
110. Kamrava, *Qatar* (2013), 35.
111. Tieleman, interview, 9 December 2012.
112. Shobrak, interview, 19 December 2012.
113. Williams, interview, 3 December 2012.
114. Ibid.
115. Emirates 24/7, 'Man' (2011).
116. Williams et al., 'Climate change' (2012), 121, 124.
117. Ibid., 121, 128.
118. Alrashidi, email, 8 January 2013.
119. Ibid., 'Curriculum' (2012).
120. Shobrak, interview, 19 December 2012.
121. Alrashidi, email, 8 January 2013.
122. Ibid., emails, 8 and 10 January 2013.
123. Ibid. et al., 'Parental cooperation' (2011), 235.
124. Ibid., email, 8 January 2013.
125. Ibid. et al., 'Integrating' (2012), 311.
126. Davidson, *After the Sheikhs* (2012), 75.

## Chapter 5    Missing Links

1. Hill et al., 'History' (1999), 21.
2. Holmes, *Hot Fossils* (1991).
3. Hill et al., 'History' (1999), 19.

4. UAE Interact, 'Exhibition' (2005).
5. Rathbone, *Abu Dhabi* (1992).
6. Kjærgaard, 'Hurrah' (2011), 83, 86.
7. Rathbone, *Abu Dhabi* (1992).
8. Sepkoski, *Rereading* (2012).
9. Gallagher, 'Reflections' (1999), 25.
10. Gallagher, 'Reflections' (1999), 25.
11. Krupp, interview, 14 January 2013.
12. Kjærgaard, 'The Fossil Trade' (2012).
13. Luomi, *The Gulf Monarchies* (2012).
14. Lipps, 'What' (1981).
15. Hill et al., 'Before Archaeology Dhabi' (2012), 22.
16. Hamilton, 'Report' (1976), [1].
17. Powers et al., *Geology of the Arabian Peninsula* (1966).
18. Glennie and Evamy, 'Dikaka' (1968).
19. Hamilton, 'Report' (1976), [1].
20. Ibid., 'Fossil Mammals' (1975).
21. Ibid., 'Report' (1976), [2].
22. Ibid.
23. Dell'Oro, Letter (1975).
24. Hamilton, Letter (1976).
25. Ball, Letter (1977).
26. Hamilton, Letter (1976), [2].
27. Ball, Letter (1978).
28. Hamilton, Memorandum (1975).
29. Kadhi, Letter (1977).
30. Hamilton, Memorandum (1978).
31. Ibid. et al., 'Fauna' (1978). Andrews et al., 'Dryopithecines (1978).
32. Andrews, Letter (1978).
33. Whybrow, 'Memorandum' (1981).
34. Masry, Letter (1987).
35. Whybrow, 'Brains' (2000), 77, 80. Whybrow and McClure, 'Fossil mangrove roots' (1981), 224.
36. Ibid. and Bassiouni, 'The Arabian Miocene' (1986). Al-Kobaisi, 'Joint research project' (1983).
37. Yasin Al Tikriti, interview, 16 December 2012.
38. Al Tikriti, 'The Impact' (2005), 11.
39. Ibid., 12.
40. Hill et al., 'History' (1999), 15.
41. Yasin Al Tikriti, interview, 16 December 2012.
42. Hill et al., 'History' (1999), 16.
43. Al Tikriti, interview, 16 December 2012.
44. Hill and Ward, 'Origin' (1988), 49.
45. Whybrow et al., 'Late Miocene Primate Fauna' (1990).

46. Ibid., 'Miocene Fossils' (1991), 4.
47. Hellyer and Aspinall, 'Researching' (2005), 18.
48. *Tribulus* 1, no. 1 (1991), [i].
49. Luomi, *The Gulf Monarchies* (2012), 107. See also Hoffman, *From Heresy* (2001), 87–106.
50. Ouis, 'Greening' (2002), 342.
51. Whybrow et al., 'Miocene Fossils' (1991), 6.
52. Whybrow, 'Initiative' (1990)
53. Ibid. and Hill, *Fossil Vertebrates* (1999), xvii.
54. Ibid. et al., 'The Fossil Record' (2005), 85.
55. Yapp, 'Adventurous oilman' (2002).
56. Brown, 'Geology' (1991), 30.
57. ADCO, 'Down Memory Lane' (2009), 11.
58. Holmes, *Hot Fossils* (1991).
59. Whybrow and Hill, 'International Seminar' [1994], 3.
60. Ibid., *Fossil Vertebrates* (1999), xvii.
61. Bibi, interview, 5 December 2012.
62. Ibid.
63. Ibid.
64. Pain, 'Breakthrough' (2009).
65. Bibi, interview, 5 December 2012.
66. Pain, 'Breakthrough' (2009).
67. Bibi, interview, 5 December 2012.
68. Pain, 'Breakthrough' (2009).
69. Davidson, *Abu Dhabi* (2009), 98.
70. Pain, 'Breakthrough' (2009).
71. Bibi, interview, 5 December 2012.
72. Ibid.
73. Higgs et al., 'A Late Miocene Proboscidean Trackway' (2003).
74. Bibi et al., 'Early evidence' (2012), 1.
75. Todorova, 'Fossils' (2010).
76. Bibi, 'Ancestors Discovered' (2008).
77. Ibid., 'Faysal' (2012), 7.
78. Ministry of Oil, 'Brief History' (2012).
79. Bourrelier and Lespine, 'Les opérations' (2008).
80. Herbert Thomas et al., 'The Lower Miocene Fauna' (1981), 110.
81. Ibid.
82. Sen, interview, 8 February 2013.
83. Bourrelier and Lespine, 'Les opérations' (2008).
84. Sen, interview, 8 February 2013.
85. Al-Sulaimani, interview, 5 February 2013.
86. Sen, interview, 8 February 2013.
87. Herbert Thomas et al., 'Early Oligocene Vertebrates' (1992); 'Découverte' (1989).
88. New Scientist, 'Ancient primates' (1988).

89. Sen, interview, 8 February 2013.
90. Al-Sulaimani, interview, 5 February 2013.
91. Herbert Thomas et al., 'Découverte' (1989), 101.
92. Ibid., 'Découverte' (1988), 823.
93. Ibid., *L'Homme* (1994), 62–4.
94. Thomas and Senut, *Les Primates* (2001), 63–78.
95. Auzias and Labourdette, *Oman* (2012), 158.
96. Sen, interview, 8 February 2013.
97. Ministry of Petroleum and Minerals and BRGM, 'Geological map' (1991; 1993).
98. Gingerich et al., 'Origin' (2001), 2239–40.
99. Saudi Geological Survey, 'Who' (2013).
100. Thisse et al., 'The Red Sea' (1983), 8.
101. Zalmout, interview, 15 September 2014. See also Kear, 'Late Cretaceous (Campanian–Maastrichtian) marine reptiles' (2008).
102. Freeth, *A piece* (2010).
103. Zalmout et al., 'New Oligocene primate' (2010), 364
104. Bhanoo, 'A Discovery' (2010).
105. Zalmout et al., 'New Oligocene primate' (2010).
106. National Science Foundation, 'New Primate Fossil' (2010).
107. Laursen, 'Fossil Skull' (2010).
108. Ibid.
109. Freeth, *A piece* (2010); *Qiṭʿah* (2010).
110. Ibid., *Qiṭʿah* (2010).
111. Laursen, 'Fossil Skull' (2010).
112. Al-ʿAṭṭās, 'Fī iktishāf' (2012). BBC, 'Saʿdān' (2010). Bhanoo, 'A Discovery' (2010). Al-Ruways, 'Iktishāf' (2010). Al-Shammarī, 'Iktishāf' (2010). Quraybī, 'Iktishāf' (2010).
113. Freeth, *Qiṭʿah* (2010);
114. Alagaili et al., 'Middle East Respiratory Syndrome' (2014).
115. Qubaysī, 'Al-jadd' (2010).
116. Freeth, *A piece* (2010).
117. Ibid., *Qiṭʿah* (2010).
118. Qubaysī, 'Al-jadd' (2010).
119. Edis, 'Cloning' (1999).
120. Brian Thomas, 'Elephant Secrets' (2009).
121. Bibi, interview, 5 December 2012.
122. Ibid.
123. Ibid.
124. Riexinger, 'Propagating' (2008), 103.
125. Enserink, 'In Europe's Mailbag' (2007). Dean, 'Islamic Creationist' (2007).
126. Raha, 'Why' (2003).
127. Al-ʿAwaḍī, 'Taḥrīr' (2008).
128. Khaleej Times, 'Turkish writer' (2007).

129. Vlaardingerbroek and Hachem-El-Masri, 'The Status' (2006), 152.
130. Al Najami, 'Debating' (2007).
131. Ibid. The astronomer Nidhal Guessoum of the American University of Sharjah also protested against the decision to move evolution from secondary schools. Guessoum, *Islam's Quantum Question* (2011), 319–20.
132. Qabbani, 'Lecturer' (2011).
133. Ibid.
134. Abdul Ghafour, 'Harun Yahya' (2008).
135. Qutb, 'Blessings' (2003).
136. Ibid., 'Genes' (2007).

## Chapter 6    Rentier Science

1. Toby Jones, *Desert Kingdom* (2010).
2. Davis, 'Environmentalism' (2000).
3. On Qatari high modernism, see Kamrava, *Qatar* (2013).
4. Marwa Elshakry, *Reading Darwin* (2013). Andrew Jones, *Developmental Fairy Tales* (2011).
5. Toby Jones, *Desert Kingdom* (2010), 54.
6. Drayton, *Nature's Government* (2000).
7. Gray, *A Theory* (2011).
8. Qatari student, personal communication, 21 September 2014.
9. On the Gulf states' tolerance, see Foley, *The Arab Gulf States* (2010).
10. Hirschler, *Medieval Arabic Historiography* (2006).
11. Edis, *An Illusion* (2007).
12. Sivasundaram, *Nature* (2005). Elshakry, 'The Gospel' (2007).
13. Hertog, *Princes* (2010), 5. See also Hertog, 'The Sociology' (2010).
14. Ibid., 'Lean' (2011), 21.
15. Hanafi, 'University systems' (2011).
16. On the private sector, see Hertog, 'State' (2013).
17. Hertog, *Princes* (2010), 249.
18. On Islamic social institutions, see Janine Clark, 'Social Movement Theory' (2004).
19. Tieleman and Williams, 'The Role' (1999). Williams and Tieleman, 'Flexibility' (2000).
20. Alrashidi et al., 'Parental cooperation' (2011); 'Integrating spatial data' (2012).
21. On the central position of the royal family in the Saudi hub-and-spoke system, see Hertog, *Princes* (2010), 58.
22. Mitchell, *Rule* (2002). More recently, Mitchell, referring to Michel Callon, challenged the distinction between experts and laypersons. Timothy, *Carbon Democracy* (2013), 240.
23. Matthiesen, *Sectarian Gulf* (2013).
24. Foley, *The Arab Gulf States* (2010), 233.
25. Fergany, 'Science' (2000), 77.

26. United Nations Development Programme, *The Arab Human Development Report* (2003), 4.

27. Waast and Rossi, 'Scientific Production' (2010), 342, 361, 347. Other scholars too placed the contemporary Arab world at the periphery of global science: Altbach, 'Centers' (2003). Donn and Al Manthri, *Globalization* (2010).

28. El Alami et al., 'International Scientific Collaboration' (1992), 262–3.

29. Callon and Rabeharisoa, 'Research' (2003).

30. Zahlan, 'Planning science' *Nature* 283 (1980), 239.

31. Willoughby, *Let a Thousand Models Bloom* (2008), 6.

32. Wang, 'Transnational Science' (2010), 375. On the transnational history of science, see also Vleuten, 'Toward' (2008), and Turchetti et al., 'Introduction' (2012).

33. E.g., Babar, 'The Cost' (2014).

34. Krupp, 'Curriculum' (2012).

35. Jones, *Desert Kingdom* (2010), 241–4. Kéchichian, *Legal and Political Reforms* (2013), 197.

36. Cittadino, *Nature* (1990).

37. Al-Rasheed, *Transnational Connections* (2005); *Kingdom* (2008). Dresch and Piscatori, *Monarchies* (2005).

38. King Saud University, 'KSU Mammals Research Chair' (2011).

39. Hamdan, 'Gulf Council' (2011).

40. F. Lacroix et al., 'The Emirates Center' (2003).

41. Van Heezik, interview, 27 September 2012.

42. F. Lacroix et al., 'The Emirates Center' (2003).

43. Chargé et al., 'Male health status' (2010), 849.

44. Davidson, *The Persian Gulf* (2010).

45. Ibid., *Persian Gulf–Pacific Asia* (2010), 24.

46. Arab News, 'KAUST's First President' (2008).

47. KACST, *The Genome of the Arabian Camel is Fully Sequenced in KSA (KACST & BGI) Part 1* (2010).

48. KACST, *The Genome of the Arabian Camel is Fully Sequenced in KSA (KACST & BGI) Part 2* (2010).

49. Mahmoud, 'Unravelling' (2010).

50. Ibid., 'Unravelling' (2010).

51. Joint Center for Genomic Research, 'Research Work' (2010).

52. Fang, 'A Complete Sequence' (2012). Al-Dous et al., '*De novo*' (2011),

53. Ibid., 'A Complete Sequence' (2012).

54. Krupp, interview, 14 January 2013.

55. Lawler, 'In search' (2014), 997.

# BIBLIOGRAPHY

This bibliography includes only sources that I have cited in my book. I arranged it alphabetically, ignoring diacritical marks but considering all forms of Arabic articles (al-, el-, etc.).

## Interviews

Abu Bakir Bagader, Berlin, 13 June 2013
Andreas Buerkert, via telephone, 7 January 2013
Annette Patzelt, via telephone, 15 January 2013
Faysal Bibi, Berlin, 5 December 2012
Friedhelm Krupp, via telephone, 14 January 2013, and in Doha, 5 October 2013
Georg Schwede, via telephone, 21 January 2013
Harald Kürschner, Berlin, 10 October 2012
Heinz Gaube, via telephone, 2 February 2013
Holger Schulz, via telephone, 9 January 2013
Irene B. Tieleman, via telephone, 9 December 2012
Iyad S. Zalmout, via Skype, 15 September 2014
Jack Roger, via email, February 2013
Joseph B. Williams, via telephone, 3 December 2012
Karl Hammer, via telephone, 9 January 2013
Martin Fisher, via Skype, 17 January 2013
Michael J. Delany, via telephone, 29 January 2013
Mohammed Shobrak, via telephone, 19 December 2012
Monif Alrashidi, via email, January 2013
Philip J. Seddon, via email and Skype, December 2012
Sevket Sen, via email and telephone, February 2013
Shahina A. Ghazanfar, London, 27 September 2012
Shaukat Ali Chaudhary, via email and telephone, January 2013
Sheila Collenette, via telephone, 20 and 27 January 2013

Stéphane Ostrowski, via email, December 2012
Thomas A. Cope, London, 20 September 2012
Walid Yasin Al Tikriti, via telephone, 16 December 2012
Wolfgang Frey, Berlin, 24 October 2012
Yolanda M. van Heezik, via Skype, 27 December 2012
Zaher Al-Sulaimani, via email and telephone, February 2013

# Primary sources

Abdel Bari, Ekhlas. 'The Flora of the State of Qatar: Its History and Present-Day Status'. *Qatar University Science Journal* 25 (2005): 113–18.

Abdul Ghafour, P.K. 'Harun Yahya: Win over Darwinism'. *Arab News*, 23 November 2008. http://www.arabnews.com/node/318407.

Abu-Zinada, Abdulaziz, Kushal Habibi, and Roland Seitre. 'The Arabian Oryx Programme in Saudi Arabia'. In *The Conservation and Biology of Desert Antelopes*, edited by Alexandra Dixon and David Jones, 41–6. London: Croom Helm, 1988.

Abū Zinādah, ʿAbd al-ʿAzīz [cf. Abdulaziz Abuzinada]. 'Liqāʾ maʿa al-shābb al-duktūr (ʿAbd al-ʿAzīz Abū Zinādah)'. *Al-Jazīrah*. 1 January 1973.

———. *Al-masīratān: al-jāmiʿah wa-al-ḥayāt al-fiṭrīyah*. Al-Riyāḍ: ʿAbd al-ʿAzīz Abū Zinādah 2011.

Abuzinada, Abdulaziz [cf. ʿAbd al-ʿAzīz Abū Zinādah]. 'Captive Breeding and Reintroduction of Native Wildlife in Saudi Arabia'. *Species* 24 (1995): 21–2.

———. 'Restoration of desert ecosystems through wildlife management – The Saudi Arabian experience'. In *Sustainable Development in Arid Zones*, edited by Samira Omar, Raafat Misak, Dhari Al-Ajmi, and Nader Al-Awadhi, 2: 641–8. Rotterdam: Balkema, 1998.

———. 'The role of protected areas in conserving biological diversity in the kingdom of Saudi Arabia'. *Journal of Arid Environments* 54 (2003): 39–45.

ADCO. 'Down Memory Lane'. *Al Waha* 27, No. 1 (2009): 11.

Ady, J. 'The National Wildlife Research Center, at Taʾif'. *Journal of the Saudi Arabian Natural History Society* 3, No. 7 (1997): 1–2.

Alagaili, Abdulaziz, Thomas Briese, Nischay Mishra, Vishal Kapoor, Stephen Sameroff, Emmie de Wit, Vincent Munster, et al. 'Middle East Respiratory Syndrome Coronavirus Infection in Dromedary Camels in Saudi Arabia'. *mBio* 5, No. 2 (2014).

AlRashidi, Monif. 'Curriculum Vitae'. *University of Haʾil*, 2012. http://faculty.uoh.edu.sa/m.alrashidi/Curriculum%20Vitae%20%28eng%29.htm.

———, András Kosztolányi, Mohammed Shobrak, Clemens Küpper, and Tamás Székely. 'Parental cooperation in an extreme hot environment: natural behaviour and experimental evidence'. *Animal Behaviour* 82, No. 2 (2011): 235–43.

———, Mohammed Shobrak, Mohammed Al-Eissa, and Tamás Székely. 'Integrating spatial data and shorebird nesting locations to predict the potential future impact of global warming on coastal habitats: A case study on Farasan Islands, Saudi Arabia'. *Saudi Journal of Biological Sciences* 19 (2012): 311–15.

Andrews, Peter. Letter to Herbert Thomas, 17 March 1978. DF PAL/140/9/167. NHM Archives.

———, W.R. Hamilton, and P.J. Whybrow. 'Dryopithecines from the Miocene of Saudi Arabia'. *Nature* 274, No. 5668 (1978): 249–51.

Arab News. 'KAUST's First President Has Impressive Credentials'. *Arab News*. 14 January 2008. http://www.arabnews.com/node/307705.

Arabian Wildlife. 'In Quest of the Baboon'. *Arabian Wildlife* 3, No. 1 (1997).

Aramco World. 'A Zoo is looking ... and ... listening'. *Aramco World* 12, No. 9 (1961): 22–4.

———. 'Portrait of an Editor'. *Aramco World* 15, No. 1 (1964): 2–7.

Asmodé, Jean-François. 'Taïf: Centre de recherche sur la faune sauvage'. *Le Courrier de la Nature* 106 (1986): 20–6.

———, and Abdul Rahman Khoja. 'Arabian Oryx Captive Breeding and Reintroduction in Saudi Arabia'. Talk given at the Aridland Antilope Workshop, San Antonio, Texas, USA, 14–15 September 1989.

Aspinall, Simon. 'Environmental Development and Protection in the UAE'. In *United Arab Emirates: A New Perspective*, edited by Ibrahim Al Abed and Peter Hellyer, 277–304. London: Trident, 2001.

———, and Philip Hockey. 'The Indian Ocean's Crab-loving Plover'. *Arabian Wildlife* 3, No. 1 (1997): 32–5.

Al-ʿAṭṭās, Ḥāmid ʿUmar. 'Fī iktishāf ʿilmī jadīd. Saʿdān al-Ḥijāz. yaẓhar fī al-suʿūdīyah baʿd 29 milyūn sanah'. *Al-Qāfilah* 61, No. 3 (2012): 42–5.

Al-ʿAwaḍī, Muḥammad. 'Taḥrīr al-ʿalmānīyīn min kawkab al-qirdah'. *Al-Raʾy*, 1 December 2008. http://www.alraimedia.com/Article.aspx?id=96040.

Auzias, Dominique, and Jean-Paul Labourdette. *Oman 2013–2014*. Petit Futé, 2012.

Aydin, Cemil. 'Introductory Essay'. *Beyond Golden Age and Decline: The Legacy of Muslim Societies in Global Modernity*, 2010. http://www.muslimmodernities.org/intro-essay-cemil-aydin.

Bagader, Abubaker, Abdullatif El-Sabbagh, Mohamad Al-Glayand, Mawil Samarrai, and Othman Llewellyn. *Environmental Protection in Islam*. Second Revised Edition. Gland: IUCN, 1994.

Baierle, Heinz Ullrich, Abdullah El-Sheikh, and Wolfgang Frey. *Vegetation und Flora im mittleren Saudi-Arabien (aṭ-Ṭāʾif – ar-Riyāḍ)*. Wiesbaden: Reichert, 1985.

Ball, H.W. Letter to A. Kadhi, 22 August 1977. DF DIR/933/16/9. NHM Archives.

———. Letter to T.A. Craig Cameron, 6 January 1978. DF DIR/933/16/9. NHM Archives.

Banaja, A. 'Introduction'. *Fauna of Saudi Arabia* 3 (1981): 2.

Bärmann, Eva Verena, Saskia Börner, Dirk Erpenbeck, Gertrud Elisabeth Rössner, Christiana Hebel, and Gert Wörheide. 'The curious case of *Gazella arabica*'. *Mammalian Biology* 78, No. 3 (2013): 220–5.

———. *Plants in the Deserts of the Middle East*. Berlin: Springer, 2001.

Bāshumayyil, Muḥammad Aḥmad. *Al-Islām wa-naẓarīyat Dārwīn*. 1964.

Basson, Philip, John Burchard, John Hardy, and Andrew Price. *Biotopes of the Western Arabian Gulf: Marine Life and Environments of Saudi Arabia*. Dhahran: Aramco, 1977.

Al-Batānūnī, Kamāl al-Dīn [cf. K.H. Batanouny]. *Nabātāt fī aḥādīth al-rasūl*. Al-Dawhah: Idārah Ihyāʾ al-Turāth al-Islāmī, 1986.

Batanouny, K.H. [cf. Kamāl al-Dīn al-Batānūnī]. *Natural History of Saudi Arabia: A Bibliography*. KAU, 1978.

————. *Ecology and Flora of Qatar*. Oxford: Alden, 1981.

Bates, Paul. 'Desert Specialists: Arabia's elegant mice'. *Arabian Wildlife* 2, No. 3 (1995): 26–7.

BBC. 'Qatari religious leader bans Pokemon'. *BBC*, 3 April 2001. http://news.bbc. co.uk/2/hi/middle_east/1258633.stm.

————. '"Saʿdān al-Ḥijāz" qad yakūn al-ḥalqah al-mafqūdah fī taṭawwur al-qirdah wa-al-insān'. *BBC Arabic*, 15 July 2010. http://www.bbc.co.uk/arabic/ scienceandtech/2010/07/100714_fossil_missing_link.shtml.

Bhanoo, Sindya N. 'A Discovery Could Help Date Monkey-Ape Split'. *The New York Times*, 19 July 2010. http://www.nytimes.com/2010/07/20/science/20obmonkey. html.

Bibi, Faysal. 'Ancestors Discovered: Paleontology and the Evolution of Life'. *American University of Beirut*, 2008. http://www.aub.edu.lb/fas/ampl/ Documents/files/lectures/BibiAbstract.pdf.

————. 'Faysal Bibi'. *American Museum of Natural History*, 2012. http://www.amnh. org/content/download/56905/915455/file/bibi_cv2012.pdf.

————, Brian Kraatz, Nathan Craig, Mark Beech, Mathieu Schuster, and Andrew Hill. 'Early evidence for complex social structure in Proboscidea from a late Miocene trackway site in the United Arab Emirates'. *Biology Letters* 8, No. 4 (2012): 670–3.

Bin Faisal, Saud [cf. Saud Al-Faisal]. 'Preface'. *Fauna of Saudi Arabia* 13 (1993): 1.

Biquand, Sylvain, Véronique Biquand-Guyot, Ahmed Boug, and Jean-Pierre Gautier. 'The Distribution of *Papio hamadryas* in Saudi Arabia: Ecological Correlates and Human Influence'. *International Journal of Primatology* 13, No. 3 (1992): 223–43.

————. 'Group Composition in Wild and Commensal Hamadryas Baboons: A Comparative Study in Saudi Arabia'. *International Journal of Primatology* 13, No. 5 (1992): 533–43.

————, S., A. Boug, V. Biquand-Guyot, and J.-P. Gautier. 'Management of Commensal Baboons in Saudi Arabia'. *Revue d'Ecologie – La Terre et La Vie* 49 (1994): 213–22.

Boulos, Loutfy. *The Weed Flora of Kuwait*. Kuwait: Kuwait University, 1988.

Breakspear, Elizabeth. Letter to J.H. Price. 'Refereeing of applications for research grants from the University of Kuwait', 7 June 1984. DF421/32/7/12. NHM Archives.

Brown, Bish. 'Geology and Palaeontology'. *Tribulus* 1, No. 1 (1991): 30.

Bucaille, Maurice. *La Bible, le Coran et la Science: Les Écritures Saintes examinées à la lumière des connaissances modernes*. Paris: Seghers, 1976.

Buerkert, Andreas. 'The Project'. *Transformation processes in oasis settlements of Oman*, 2012. http://www.agrar.uni-kassel.de/ink/oman/sites/project.htm.

————, and Jens Gebauer, eds. *Agrobiodiversity and genetic erosion: Contributions in Honor of Prof. Dr. Karl Hammer*. Kassel: Kassel University Press, 2009.

————, and Eva Schlecht, eds. *Oases of Oman: Livelihood Systems at the Crossroads*. Second expanded edition. Muscat: Al Roya, 2010.

Burchard, John. Letter to P.F.S. Cornelius, 9 December 1980. DF253/192/3. NHM Archives.

————. Letter to P.F.S. Cornelius, 11 March 1981. DF253/192/3. NHM Archives.

Büttiker, W. 'Further Notes on the Zoological Survey of Saudi Arabia'. *Fauna of Saudi Arabia* 3 (1981): 5–24.

————, and F. Krupp. 'Introduction'. *Fauna of Saudi Arabia* 6 (1984): 2–3.

————. 'Introduction'. *Fauna of Saudi Arabia* 7 (1985): 2–3.

————. 'Introduction'. *Fauna of Saudi Arabia* 10 (1989): 2.

————. 'Introduction'. *Fauna of Saudi Arabia* 14 (1994): 2.

Carey, Bjorn. 'Desert Gazelles Found to Shrink Their Hearts to Save Water'. *Fox News*, 12 June 2006. http://www.foxnews.com/story/0,2933,198928,00.html.

Carter, Humphrey. 'Some Plants of the Zor Hills, Koweit, Arabia'. *Records of the Botanical Survey of India* 6, No. 6 (1917): 175–306.

Chargé, Rémi, Michel Saint Jalme, Frédéric Lacroix, Adeline Cadet, and Gabriele Sorci. 'Male health status, signalled by courtship display, reveals ejaculate quality and hatching success in a lekking species'. *Journal of Animal Ecology* 79, No. 4 (2010): 843–50.

Chaudhary, Shaukat. *Grasses of Saudi Arabia*. Riyadh: Ministry of Agriculture and Water, 1989.

————. *Flora of the Kingdom of Saudi Arabia Illustrated*. Vol. 1. Riyadh: Ministry of Agriculture and Water, 1999.

————, and Muhammad Akram. *Weeds of Saudi Arabia and the Arabian Peninsula*. Riyadh: Ministry of Agriculture and Water, 1987.

————, and Abdul Aziz Al-Jowaid. *Vegetation of the Kingdom of Saudi Arabia*. Riyadh: Ministry of Agriculture and Water, 1999.

———— and Mouafaq Zawawi. *A Manual of Weeds of Central and Eastern Saudi Arabia*. Riyadh: Ministry of Agriculture and Water, 1983.

Clark, Arthur. 'In the Lions' Den'. *Aramco World* 36, No. 1 (1985): 36–40.

————. 'A Capital Zoo in Riyadh'. *Aramco World* 41, No. 2 (1990): 36–40.

Collenette, Sheila. *Wildflowers of Saudi Arabia*. Riyadh: National Commission for Wildlife Conservation and Development, 1999.

Comptroller General of the United States. *The U.S.–Saudi Arabian Joint Commission On Economic Cooperation*. Washington: United States General Accounting Office, 22 March 1979.

Cope, Thomas. *A Key to the Grasses of the Arabian Peninsula*. Riyadh: Arab Bureau of Education for the Gulf States, 1985.

Daly, Ralph. '"And what," said the Sultan, "shall we do about the oryx?"' *Species Survival Commission Newsletter* 5 (1985): 25–7.

Darwin, Charles. *On the Origin of Species by Means of Natural Selection: Or, The Preservation of Favoured Races in the Struggle for Life*. London: John Murray, 1859.

Dell'Oro, Walter. Letter to W. Roger Hamilton, 4 March 1975. DF DIR/933/16/9. NHM Archives.

Dickson, David. 'Princeton and Saudi universities sign life sciences agreement'. *Nature* 284, No. 5752 (1980): 112.

Dickson, V. 'Plants of Kuwait, North-East Arabia'. *Journal of the Bombay Natural History Society* 40, No. 3 (1938): 528–38.

Al-Dous, Eman, Binu George, Maryam Al-Mahmoud, Moneera Al-Jaber, Hao Wang, Yasmeen Salameh, Eman Al-Azwani, et al. 'De novo genome sequencing and comparative genomics of date palm (*Phoenix dactylifera*)'. *Nature Biotechnology* 29, No. 6 (2011): 521–7.

Emirates 24/7. 'Man banned for showing dog with Saudi dress'. *Emirates 24/7*, 12 May 2011. http://www.emirates247.com/news/man-banned-for-showing-dog-with-saudi-dress-2011-05-12-1.391859.

Enserink, Martin. 'In Europe's Mailbag: A Glossy Attack on Evolution'. *Science* 315, No. 5814 (2007): 925.

Al-Faisal, Saud [cf. Saud bin Faisal]. 'Preface'. *Fauna of Saudi Arabia* 9 (1988): 1.

Fang, Y., H. Wu, T. Zhang, M. Yang, Y. Yin, L. Pan, X. Yu, X. Zhang, S. Hu, and I.S. Al-Mssallem. 'A Complete Sequence and Transcriptomic Analyses of Date Palm (*Phoenix Dactylifera L.*) Mitochondrial Genome'. *PloS One* 7, No. 5 (2012).

Al Farhan, Ahmed H., Ibrahim Aldjain, Jacob Thomas, Anthony Miller, Sabina Knees, Owen Lewellyn, and Ali Akram. 'Botanic Gardens in the Arabian Peninsula'. *Sibbaldia* 6 (2008): 189–203.

Fauna & Flora International. 'Bahrain to the Rescue'. *Oryx* 15, No. 3 (1980): 231.

Fisher, Martin, Shahina Ghazanfar, and Andrew Spalton, eds. *The Natural History of Oman*. Leiden: Backhuys, 1999.

Flamand, Jacques. 'Arabian Oryx: Run Wild, Run Free!' *Arabian Wildlife* 1, No. 1 (1994): 12–13.

Freeth, Martin. *A piece in the monkey puzzle*. Nature Video, 2010. http://www.youtube.com/watch?v=r2-RkQJ-3xo.

———. *Qiṭʿah ukhrá li-ḥall laghz nushūʾ raʾīsīyāt*. Nature Video, 2010. http://www.youtube.com/watch?v=fOBlJLiNKkM.

Frey, Wolfgang. 'Neue Vorstellungen über die Verwandtschaftsgruppen und die Stammesgeschichte der Laubmose'. In *Beiträge zur Biologie der niederen Pflanzen: Systematik, Stammesgeschichte, Ökologie*, edited by W. Frey, H. Hurka, and F. Oberwinkler, 117–39. Stuttgart: Gustav Fischer, 1977.

———. 'A tribute to Harald Kürschner'. In *Bryophyte Systematics, Phytodiversity, Phytosociology and Ecology*, edited by Wolfgang Frey, 9–24. Stuttgart: Cramer, 2010.

———, and Harald Kürschner. 'The first records of bryophytes from Saudi Arabia'. *Lindbergia* 8, No. 3 (1982): 157–60.

———. *Tübingen Atlas of the Near and Middle East: Final Report of the Section Botany, Supporting Period 1974 – 1985*. Berlin: Freie Universität Berlin, 1985.

———. 'Masqaṭ area (Oman). Remnants of vegetation in an urban habitat'. In *Contributions to the Vegetation of Southwest Asia*, edited by Harald Kürschner, 201–21. Wiesbaden: Reichert, 1986.

———. 'Wüstenmoose: Anpassungen und Überlebensstrategien im täglichen Kampf mit der Sonne'. *Biologie in unserer Zeit* 28, No. 4 (1998): 231–40.

Frey, Wolfgang, and Wilfried Probst. *Vegetation und Flora des Zentralen Hindūkuš (Afghanistan)*. Wiesbaden: Reichert, 1978.

Al-Gain, A. 'Preface'. *Fauna of Saudi Arabia* 7 (1985) 7: 1.

Gallagher, M.D. Letter to I.C.J. Galbraith, 23 June 1977. DF230-122-116. NHM Archives.

———. Letter to I.C.J. Galbraith, 4 July 1979. DF230-122-116. NHM Archives.

———. 'Reflections'. In *The Natural History of Oman*, edited by Martin Fisher, Shahina Ghazanfar, and Andrew Spalton, 23–8. Leiden: Backhuys, 1999.

Garcia, Diane. 'Gazelles Get That Shrinking Feeling'. *ScienceNOW*, 29 June 2006. http://news.sciencemag.org/sciencenow/2006/06/29-02.html.

Gautier, Jean-Pierre, and Sylvain Biquand. 'Primate Commensalism'. *Revue d'Écologie – La Terre et la Vie* 49, No. 3 (1994): 210–12.

Gebauer, Jens, Sulaiman Al Khanjari, Iqrar Khan, Andreas Buerkert, and Karl Hammer. 'Plant Genetic Resources in Oman – Evidence for Millennia of

Cultural Exchange in the Middle East'. In *Oases of Oman: Livelihood Systems at the Crossroads*, edited by Andreas Buerkert and Eva Schlecht, 28–33. Second expanded edition. Muscat: Al Roya, 2010.

Ghazanfar, Shahina A. 'CV'. Word document sent to me by the author, 10 September 2012.

Gießner, Klaus. 'Das Afrika-Kartenwerk'. *Africa Spectrum* 4, No. 2 (1969): 30–3.

Gingerich, Philip, Munir ul Haq, Iyad Zalmout, Intizar Khan, and Sadiq Malkani. 'Origin of Whales from Early Artiodactyls: Hands and Feet of Eocene Protocetidae from Pakistan'. *Science* 293, No. 5538 (2001): 2239–42.

Glennie, K.W., and B.D. Evamy. 'Dikaka: Plants and Plant-root Structures Associated with Aeolian Sand'. *Palaeogeography, Palaeoclimatology, Palaeoecology* 4, No. 2 (1968): 77–87.

Government of Dubai Media Office. 'Dubai International Holy Quran Award'. *His Highness Sheikh Mohammed Bin Rashid Al Maktoum*, 2013. http://www.shei khmohammed.com/vgn-ext-templating/v/index.jsp?vgnextoid=5ec1e1b33a3 4110VgnVCM1000003f140a0aRCRD&vgnextchannel=0561fd70bdc043 10VgnVCM1000004d64a8c0RCRD&vgnextfmt=default&date=106016493 7033.

Greth, Arnaud, Bruno Andral, Hermann Gerbermann, Marc Vassart, Helga Gerlach, and Frederic Launay. 'Chlamydiosis in a Captive Group of Houbara Bustards (Chlamydotis undulata)'. *Avian Diseases* (1993): 1117–20.

Greth, A., J.R.B. Flamand, and A. Delhomme. 'An outbreak of tuberculosis in a captive herd of Arabian oryx (*Oryx leucoryx*): management'. *The Veterinary Record* 134, No. 7 (1994): 165–7.

Greth, Arnaud, and Chris Magin. 'Final Discussion, Guidelines and Recommendations of the Arabian Gazelle Working Group'. In *Conservation of Arabian Gazelles*, edited by Arnaud Greth, Chris Magin, and Marc Ancrenaz, 159–68. Riyadh: NCWCD, 1996.

———, Georg Schwede, Marc Vassart, and Laurent Granjon. 'The Controversial Subject of Gazelle Subspecies in Saudi Arabia and Its Significance for Reintroduction Programmes'. In *Conservation of Arabian Gazelles*, edited by Arnaud Greth, Chris Magin, and Marc Ancrenaz, 51–68. Riyadh: NCWCD, 1996.

———, and Douglas Williamson. 'Recent Developments in Gazelle Conservation and Taxonomy in Saudi Arabia'. In *Conservation of Arabian Gazelles*, edited by Arnaud Greth, Chris Magin, and Marc Ancrenaz, 1–7. Riyadh: NCWCD, 1996.

Grimwood, Ian. '"Operation Oryx": The Second Stage'. *Oryx* 7, No. 5 (1964): 223–5.

Gulf News. 'Saudis ban Pokemon as gambling, un-Islamic'. *Gulf News*, 1 April 2001. http://gulfnews.com/news/gulf/uae/general/saudis-ban-pokemon-as-gam bling-un-islamic-1.412082.

Gutberlet, Manuela. 'Jabel Al Akhdar fruit crops in danger: Expert'. *Oman Tribune*, 15 May 2007.

———. 'Oases in Al Jabel Al Akhdar can do wonders'. *Oman Tribune*, 19 May 2007.

Al-Ḥājj, Mayy. 'Al-masīratān... al-jāmiʿah wa-al-ḥayāt al-fiṭrīyah'. *Laha Magazine*, 2011. http://www.lahamag.com/Details/11563/%27%D8%A7%D9%84%D9% 85%D8%B3%D9%8A%D8%B1%D8%AA%D8%A7%D9%86..._%D8% A7%D9%84%D8%AC%D8%A7%D9%85%D8%B9%D8%A9_%D9%88% D8%A7%D9%84%D8%AD%D9%8A%D8%A7%D8%A9_%D8%A7%D9% 84%D9%81%D8%B7%D8%B1%D9%8A%D8%A9%27.

Hamdan, Sara. 'Gulf Council Reaches Out to Morocco and Jordan'. *The New York Times*, 25 May 2011. http://www.nytimes.com/2011/05/26/world/middleeast/26iht-M26-GCC.html.

Hamilton, W.R. Memorandum to the Director, 30 April 1975. DF DIR/933/16/9. NHM Archives.

———. 'Fossil Mammals from Saudi Arabia: Notes on a Collection Made by W.R. Hamilton and P.J. Whybrow,' 5 May 1975. DF DIR/933/16/9. NHM Archives.

———. Letter to E. Schysfma, 25 May 1976. DF DIR/933/16/9. NHM Archives.

———. 'Report on the Saudi Arabian Collection,' 18 July 1976. DF DIR/933/16/9. NHM Archives.

———. Memorandum to the Director, 1 February 1978. DF DIR/933/16/9. NHM Archives.

———, P.J. Whybrow, and H.A. McClure. 'Fauna of Fossil Mammals from the Miocene of Saudi Arabia'. *Nature* 274, No. 5668 (1978): 248–9.

Hammer, K., J. Gebauer, S. Al Khanjari, and A. Buerkert. 'Oman at the cross-roads of inter-regional exchange of cultivated plants'. *Genetic Resources and Crop Evolution* 56, No. 4 (2009): 547–60.

Hammond, R.L., W. Macasero, B. Flores, O.B. Mohammed, T. Wacher, and M.W. Bruford. 'Phylogenetic Reanalysis of the Saudi Gazelle and Its Implications for Conservation'. *Conservation Biology* 15, No. 4 (2001): 1123–33.

Harrison, D.L., and P.J. Bates. 'On the occurrence of the European Free-tailed Bat, *Tadarida teniotis Rafinesque*, 1814 (*Chiroptera: Molossidae*) in Saudi Arabia'. *Fauna of Saudi Arabia* 6 (1984): 551–6.

Häser, Jutta, Eike Luedeling, Eva Schlecht, and Andreas Buerkert. 'From Hunter-Gatherer Communities to Oasis Cultures: Climate Change and Human Adaptation on the Oman Peninsula'. In *Oases of Oman: Livelihood Systems at the Crossroads*, edited by Andreas Buerkert and Eva Schlecht, 16–27. Second expanded edition. Muscat: Al Roya, 2010.

Heezik, Yolanda van, and Philip Seddon. 'Born To Be Tame'. *Natural History* 110, No. 5 (2001): 58–63.

Hellyer, Peter. 'The Emirates Natural History Group'. *Arabian Wildlife* 1, No. 1 (1994): 21.

———. 'In Remembrance: A lover, and great observer, of the natural UAE'. *The National*, 4 November 2011. http://www.thenational.ae/news/uae-news/environment/in-remembrance-a-lover-and-great-observer-of-the-natural-uae.

———, and Simon Aspinall. 'Researching the Emirates'. In *The Emirates: A Natural History*, edited by Peter Hellyer and Simon Aspinall, 13–25. Trident, 2005.

Higgs, Will, Anthony Kirkham, Graham Evans, and Dan Hull. 'A Late Miocene Proboscidean Trackway from Western Abu Dhabi'. *Tribulus* 13, No. 2 (2003): 3–8.

Hill, Andrew, Faysal Bibi, Mark Beech, and Walid Yasin Al Tikriti. 'Before Archaeology: Life and Environments in the Miocene of Abu Dhabi'. In *Fifty Years of Emirates Archaeology*, edited by D.T. Potts and P. Hellyer, 21–33. Abu Dhabi: Motivate, 2012.

———, and Steven Ward. 'Origin of the Hominidae: The Record of African Large Hominoid Evolution Between 14 My and 4 My'. *American Journal of Physical Anthropology* 31 (1988): 49–83.

———, Peter Whybrow, and Walid Yasin. 'History of Palaeontological Research in the Western Region of the Emirate of Abu Dhabi, United Arab Emirates'. In

*Fossil Vertebrates of Arabia: With Emphasis on the Late Miocene Faunas, Geology, and Palaeoenvironments of the Emirate of Abu Dhabi, United Arab Emirates*, edited by Peter J. Whybrow and Andrew Hill, 15–23. New Haven: Yale University Press, 1999.

Hockey, Phil, and Simon Aspinall. 'The Crab Plover: Enigmatic Wader of the Desert Coasts'. *Africa – Birds & Birding* 1, No. 1 (1996): 60–7.

———. 'Why do Crab Plovers *Dromas ardeola* breed in colonies?' *Bulletin of the International Wader Study Group* 82 (1997): 38–42.

Holmes, Dave. *Hot Fossils from Abu Dhabi*. NHM, 1991.

Hoye, Paul. 'Return of the Oryx'. *Aramco World* 33, No. 4 (1982): 14–17.

IFHC. 'The Future of Falconry Intrinsically Linked to the Survival of the Houbara Bustard'. *IFHC*, 14 December 2011. http://www.houbarafund.org/docs/press releases/iff-press-release-eng-14-dec-2011.pdf.

Ipsos. 'Ipsos Global @dvisory: Supreme Being(s), the Afterlife and Evolution'. *Ipsos*, 25 April 2011. http://www.ipsos-na.com/news-polls/pressrelease.aspx?id=5217.

Islam, Zafar-ul, and Ahmed Boug. 'Save the Houbara.' *Hornbill* (2011): 18–21.

Jennings, Michael C. *Atlas of the Breeding Birds of Arabia*. Basle: Karger Libri, 2010.

Joint Center for Genomic Research. 'Research Work'. *KACST*, 2010. https://www.kacst.edu.sa/en/depts/jcg/researchwork/Pages/default.aspx.

Jungius, H. 'Plan to Restore Arabian Oryx in Oman'. *Oryx* 14, No. 4 (1978): 328–36.

KACST. *The Genome of the Arabian Camel Is Fully Sequenced in KSA (KACST & BGI) Part 1*, 2010. http://www.youtube.com/watch?feature=player_embedded& v=A1AoSDkQ42U.

———. *The Genome of the Arabian Camel Is Fully Sequenced in KSA (KACST & BGI) Part 2*, 2010. http://www.youtube.com/watch?v=ZB3R464o97A&feature= youtube_gdata_player.

Ba Kader, Abou Bakr, Abdul Latif Al Sabbagh, Mohamed Al Glenid, and Mouel Samarrai. *Basic Paper on the Islamic Principles for the Conservation of the Natural Environment*. Gland: IUCN; MEPA, 1983.

Kadhi, A. Letter to H.A. McClure, 19 July 1977. DF DIR/933/16/9. NHM Archives.

Kear, B.P., T.H. Rich, M.A. Ali, Y.A. Al-Mufarrih, A.H. Matiri, A.M. Masary, and Y. Attia. 'Late Cretaceous (Campanian–Maastrichtian) marine reptiles from the Adaffa Formation, NW Saudi Arabia'. *Geological Magazine* 145, No. 5 (2008): 648–54.

Khaleej Times. 'Discoverer of Abu Dhabi's fossils dies'. *Khaleej Times*, 17 February 2004. http://www.khaleejtimes.com/DisplayArticle.asp?xfile=data/theuae/ 2004/February/theuae_February316.xml&section=theuae&col=.

———. 'Turkish writer plans expo of 'Living Fossils' in Dubai'. *Khaleej Times*, 24 January 2007. http://www.khaleejtimes.com/kt-article-display-1.asp? section=theuae&xfile=data/theuae/2007/january/theuae_january773.xml.

Al Khanjari, Sulaiman. *Exploration and estimation of morphological and genetic diversity of wheat (Triticum spp.) landraces in Oman*. Kassel: Kassel University Press, 2005.

Al Khanjari, S., K. Hammer, A. Buerkert, and M.S. Röder. 'Molecular diversity of Omani wheat revealed by microsatellites: I. Tetraploid landraces'. *Genetic Resources and Crop Evolution* 54, No. 6 (2007): 1291–300.

———, A.A. Filatenko, K. Hammer, and A. Buerkert. 'Morphological spike diversity of Omani wheat'. *Genetic Resources and Crop Evolution* 55, No. 8 (2008): 1185–95.

Khattab, Ahmed, and Nabil El-Hadidi. *Results of a Botanic Expedition to Arabia in 1944-1945.* Cairo University Press, 1971.

King Saud University. 'KSU Mammals Research Chair, Zoological Society of London (ZSL) sign cooperation agreement'. *KSU*, 10 October 2011. http://enews.ksu. edu.sa/2011/10/10/ksu-mammals-chair-zoological-society-london/.

Klausewitz, W. 'Evolutionary History and Zoogeography of the Red Sea Ichthyofauna'. *Fauna of Saudi Arabia* 10 (1989): 310–37.

Al-Kobaisi, A.J. Letter to Peter J. Whybrow. 'Joint research project of Drs. P.J. Whybrow and Peter Andrews, The British Museum, and Prof. Dr. M.A. Bassiouni, Qatar University', 30 June 1983. DF PAL/140/9/167. NHM Archives.

Krupp, Friedhelm. 'Curriculum Vitae'. Document sent to me by the author, July 2012.

———, William Büttiker, Iyad Nader, and Wolfgang Schneider. 'Introduction'. *Fauna of Arabia* 21 (2006): 4–5.

———, and Volker Mahnert. 'Introduction'. *Fauna of Saudi Arabia* 15 (1996): 2–3.

———, Wolfgang Schneider, and Ragnar Kinzelbach. 'Preface'. In *Proceedings of the Symposium of the Fauna and Zoogeography of the Middle East*, edited by Friedhelm Krupp, Wolfgang Schneider, and Ragnar Kinzelbach, 7–8. Wiesbaden: Reichert, 1987.

Kummer, Hans. *In Quest of the Sacred Baboon: A Scientist's Journey.* Translated by Ann Biederman-Thorson. Princeton: Princeton University Press, 1995.

Kürschner, Harald. 'Biogeography of South-West Asian Bryophytes – With Special Emphasis on the Tropical Element'. *Turkish Journal of Botany* 32 (2008): 433–46.

———, and Shahina Ghazanfar. 'Bryophytes and Lichens'. In *Vegetation of the Arabian Peninsula*, edited by Shahina Ghazanfar and Martin Fisher, 99–124. Dordrecht: Kluwer, 1998.

Kuwait University. 'KTUH History'. *Kuwait University*, 2009. http://biosci.kuniv. edu/Hr_History.htm.

Lacroix, F., J. Seabury, M. Al Bowardi, and J. Renaud. 'The Emirates Center for Wildlife Propagation: Comprehensive Strategy to Secure Self-Sustaining Wild Populations of Houbara Bustard (*Chlamydotis undulata undulata*) in Eastern Morocco'. *Wildlife Middle East News* (2003): 60–2.

Al-Lajnah al-Dāʾimah lil-Buḥūth al-ʿIlmīyah wa-al-Iftāʾ. 'Al-taṭawwur wa-al-irtiqāʾ: naẓarīyat Dārwīn'. *Al-Riʾāsah al-ʿĀmmah lil-Buḥūth al-ʿIlmīyah wa-al-Iftāʾ.* Accessed 7 July 2014. http://www.alifta.net/fatawa/fatawaDetails.aspx? BookID=3&View=Page&PageNo=7&PageID=10478&languagename=.

———. 'Naẓarīyat al-taṭawwur wa-al-irtiqāʾ'. *Al-Riʾāsah al-ʿĀmmah lil-Buḥūth al-ʿIlmīyah wa-al-Iftāʾ.* Accessed 18 July 2014. http://alifta.com/Fatawa/ FatawaChapters.aspx?languagename=ar&View=Page&PageID=11& PageNo=1&BookID=3.

Lang, Hans-Joachim. 'Heinz Gaube lebt und forscht im Sultanat Oman'. *Schwäbisches Tagblatt Tübingen*, 18 May 2011. http://www.tagblatt.de/Home/ nachrichten/hochschule_artikel,-Heinz-Gaube-lebt-und-forscht-im-Sultanat-Oman-_arid,134575.html.

Larivière, Serge, and Philip Seddon. 'Vulpes rueppelli'. *Mammalian Species* 678 (2001): 1–5.

Launay, Frédéric. 'Conclusions and recommendations'. In *Counting Houbara Bustards*, edited by Fred Launay and Theri Bailey, 25–6. Abu Dhabi: Environmental Research and Wildlife Development Agency, 1997.

———, and Patrick Paillat. 'A Behavioural Repertoire of the Adult Houbara Bustard (*Chlamydotis undulata macqueenii*)'. *Revue d'Écologie – La Terre et La Vie* 45 (1990): 65–88.

Lewin, Tamar. 'U.S. Universities Join Saudis in Partnerships'. *The New York Times*, 6 March 2008. http://www.nytimes.com/2008/03/06/education/06partner.html.

Mahmoud, Abdulhakim. 'Unravelling the camel's genetic sequence'. *Nature Middle East*, 24 June 2010. http://www.nature.com/nmiddleeast/2010/100624/full/nmiddleeast.2010.166.html.

Mandaville, James. *Flora of Eastern Saudi Arabia*. London: Kegan Paul International, 1990.

———. *Bedouin Ethnobotany: Plant Concepts and Uses in a Desert Pastoral World*. University of Arizona Press, 2011.

De Marco, Giovanni, and Angela Dinelli. 'First contribution to the floristic knowledge of Saudi Arabia'. *Annali Di Botanica* 33 (1974): 209–36.

Masry, Abdullah. Letter to Peter Whybrow, 27 July 1987. DF PAL/140/9/167. NHM Archives.

Michaelis, Johann David. *Fragen an eine Gesellschaft Gelehrter Männer, die auf Befehl Ihro Majestät des Königes von Dännemark nach Arabien reisen*. Frankfurt: Johann Gottlieb Garbe, 1762.

Migahid, A. M. *Migahid and Hammouda's Flora of Saudi Arabia*. Riyadh University, 1978.

———, A.F. Shalaby, M.I. Ali, and A.H. Abu-Zinada, eds. *Proceedings of the First Conference on the Biological Aspects of Saud Arabia*. Riyad University Press, 1977.

Miller, A.G., and T.A. Cope. *Flora of the Arabian Peninsula and Socotra*. Vol. 1. Edinburgh: Edinburgh University Press, 1996.

Ministry of Oil. 'Brief History'. *Ministry of Oil*, 2012. http://www.moo.gov.kw/About-Us/Brief-History.aspx.

Ministry of Petroleum and Minerals, and BRGM. 'Geological map, Oman 1:100,000: NF 40-8E, Ja'alan'. [Muscat:] Ministry of Petroleum and Minerals, 1991.

———. 'Geological map, Oman 1:100,000: NE 39-12E, Mudavy'. [Muscat:] Ministry of Petroleum and Minerals, 1993.

Al Najami, Siham. 'Debating the origin of life'. *Gulf News*, 17 February 2007. http://gulfnews.com/news/gulf/uae/heritage-culture/debating-the-origin-of-life-1.161641.

Nasseef, Abdullah. 'Role of Water in Coloring Rocks: A Theogeological Approach'. In *Geology of the Arab World*, edited by Ali Sadek, 1:3–11. Cairo: Cairo University, 1992.

National Science Foundation. 'New Primate Fossil Found in Saudi Arabia'. *National Science Foundation*, 14 July 2010. http://www.nsf.gov/news/news_summ.jsp?cntn_id=117334.

National Scientific and Technical Information Center. *Iraqi aggression on Kuwait Institute for Scientific Research: effects and repercussions*. Kuwait: Center for Research and Studies on Kuwait, 1999.

New Scientist, 'Ancient Primates Lived Out of Africa', *New Scientist*, 10 March 1988.

Ostrowski, Stéphane. 'Ajustements écophysiologiques des antilopes au milieu désertique'. PhD thesis, Université Claude-Bernard – Lyon I, 2006.

Pain, Elisabeth. 'Breakthrough of the Year: A Tale of Two Paleontologists'. *Science*, 18 December 2009. doi:10.1126/science.caredit.a0900155.

Percy, Bill. 'Obituary: Ralph Hinshelwood Daly OBE (1924–2006)'. *The British-Yemeni Society*, 2006. http://www.al-bab.com/bys/obits/daley07.htm.

Du Plessis, Morné, and Joseph Williams. 'Communal Cavity Roosting in Green Woodhoopoes: Consequences for Energy Expenditure and the Seasonal Pattern of Mortality'. *Auk* 111, No. 2 (1994): 292–9.

Powers, R.W., L.F. Ramirez, C.D. Redmond, and E.L. Elberg. *Geology of the Arabian Peninsula: Sedimentary Geology of Saudi Arabia*. US Geological Survey Professional Paper 560-D. Washington: United States Government Printing Office, 1966.

Qabbani, Bana. 'Lecturer claims fossils "validate creation"'. *The National*, 11 January 2011. http://www.thenational.ae/news/uae-news/education/lecturer-claims-fossils-validate-creation.

Al-Qaraḍāwī, Yūsuf, and ʿUthmān ʿUthmān. 'Bidāyat al-khalq wa-naẓarīyat al-taṭawwur'. *Al-Qaraḍāwī*, 26 September 2011. http://www.qaradawi.net/2010-02-23-09-38-15/4/866-2011-10-13-10-36-59.html.

Qatar Digital Library. 'About the Qatar Digital Library'. *Qatar Digital Library*, 30 July 2014. http://www.qdl.qa/en/about.

Qubaysī, Kamāl. 'Al-jadd al-arḍī al-akbar "Saʿdāniyūs ḥijāzinsīs" ʿāsha qurb Makkah'. *Al-ʿArabīyah*, 15 July 2010. http://www.alarabiya.net/articles/2010/07/15/113949.html.

Quraybī, Khalīl. 'Iktishāf 'Saʿdān a-Ḥijāz' al-munqariḍ qabla 29 milyūn sanah fī al-Jamūm'. *Al-Madīnah*, 17 July 2010. http://www.al-madina.com/node/256762.

Qutb, Sayyid. 'Blessings All Around Us'. *Arab News*, 21 March 2003. http://www.arabnews.com/node/229513.

———. 'Genes, Evolution and the Making of Man'. *Arab News*, 20 July 2007. http://www.arabnews.com/node/300886.

Raha, Sonali. 'Why you should go to Sharjah Book Fair'. *Gulf News*, 10 December 2003. http://gulfnews.com/news/gulf/uae/general/why-you-should-go-to-sharjah-book-fair-1.372857.

Rathbone, John. *Abu Dhabi: the missing link*. Abu Dhabi National Oil Company and ADCO, 1992.

Renaud, Jacques. 'Jacques Renaud'. *LinkedIn*, 2013. http://www.linkedin.com/pub/jacques-renaud/19/785/b3b.

Roberts, M.B.V. *Biology: A Functional Approach*. 4th edition. Cheltenham: Nelson Thornes, 1986.

Al-Ruways, Ṣāliḥ. 'Iktishāf nawʿ jadīd min ashbāh al-saʿādīn ʿāshat qabla 29 milyūn sanah fī al-Ḥijāz'. *Al-Riyāḍ*, 16 July 2010. http://www.alriyadh.com/544285.

Saghir, Abdur-Rahman, and Shaukat Chaudhary. *Weed Control Handbook for Saudi Arabia*. Riyadh: Ministry of Agriculture and Water, 1985.

Al-Saihati, Abdul-Wahed. 'Studies in falconry and conservation of Houbara Bustard (*Chlamydotis undulata macqueenii*) in Saudi Arabia'. PhD, Union Institute, 1996.

Samour, Jaime, James Irwin-Davies, Mubarak Mohanna, and Essa Faraj. 'Conservation at Al-Areen Wildlife Park, Bahrain'. *Oryx* 23, No. 3 (1989): 142–5.

Al Saud, Abdullah Bin Abdulaziz. 'King's Speech at KAUST Inauguration Ceremony'. *KAUST*, 23 September 2009. http://www.kaust.edu.sa/speeches/KingAbdullahInauguralSpeech.html.

——, Sultan bin Abdulaziz. 'Preface'. *Courier Forschungsinstitut Senckenberg* 166 (1994): 1.

Saudi Arabian Cultural Mission to the U.S.A. *Directory of Doctoral Dissertations of Saudi Graduates from U.S. Universities (1964–2005)*. 2006.

Saudi Biological Society. *Proceedings: Second Conference on the Biological Aspects of Saudi Arabia*. Saudi Biological Society, 1978.

Saudi Geological Survey. 'Who We Are'. *Saudi Geological Survey*, 2013. http://www.sgs.org.sa/English/AboutSGS/Who_We_Are/Pages/default.aspx.

Seddon, P.J., M. Saint Jalme, Y. van Heezik, P. Paillat, P. Gaucher, and O. Combreau. 'Restoration of houbara bustard populations in Saudi Arabia: developments and future directions'. *Oryx* 29, No. 2 (1995): 136–42.

Al-Shammarī, Nahār. 'Iktishāf ʿSaʿdān al-Ḥijāz' bi-Jiddah yaʿūd ilá 30 milyūn ʿām'. *Al-Yawm*, 17 July 2010. http://www.alyaum.com/article/2774467.

El-Sheikh, Abdullah Mohammad. 'International Cooperation'. *King Saud University*, 2012. http://faculty.ksu.edu.sa/23498/Pages/WCooperation.aspx.

Shobrak, Mohammed. 'Ecology of the Lappet-faced Vulture *Torgos tracheliotus* in Saudi Arabia'. PhD thesis, University of Glasgow, 1996.

Shrek. 'Ṭālibāt yashkīn min tadrīs "naẓarīyat Dārwīn" bi-iḥdá kullīyāt al-Suʿūdīyah'. *Arab Atheists Network*, 1 April 2007. http://www.il7ad.com/smf/index.php?topic=6944.0.

Smalley, Ben. 'Dubai scholars declare Pokemon unislamic'. *Gulf News*, 9 April 2001. http://gulfnews.com/news/gulf/uae/general/dubai-scholars-declare-pokemon-unislamic-1.412648.

Spalton, Andrew. 'Ralph Hinshelwood Daly OBE'. *Species*, No. 46 (2006): 29–30.

Spalton, J.A., M.W. Lawerence, and S.A. Brend. 'Arabian oryx reintroduction in Oman: successes and setbacks'. *Oryx* 33, No. 2 (1999): 168–75.

Stanley Price, Mark R. *Animal Re-introductions: The Arabian oryx in Oman*. Cambridge: Cambridge University Press, 1989.

Talhouk, A.S. 'Introduction'. *Fauna of Saudi Arabia* 2 (1980): 1–2.

Teller, Matthew. 'Rx for Oryx'. *Saudi Aramco World* 60, No. 5 (2009): 12–23.

——. 'The happy ones?' *Saudi Aramco World* 63, No. 6 (2012): 10–16.

Temple, A.I. 'Letter to the Keeper, the Department of Botany, British Museum (Natural History)', 20 May 1950. DF425/1/15/8. NHM Archives.

Thisse, Yves, Pol Guennoc, Georges Pouit, and Zohair Nawab. 'The Red Sea: A Natural Geodynamic and Metallogenic Laboratory'. *Episodes* 6, No. 3 (1983): 3–8.

Al Tikriti, Walid Yasin. 'The Impact of Archaeology on the Paleontology of the Western Region of Abu Dhabi: the History of Palaeontological Research'. In *Abu Dhabi – 8 Million Years Ago: Late Miocene Fossils from the Western Region*, edited by Mark Beech and Peter Hellyer, 11–14. Abu Dhabi: ADIAS, 2005.

Time. 'A Kingdom for the Oryx'. *Time* 82, No. 2 (1963): 54.

Thomas, Brian. 'Elephant Secrets under Middle East Sands'. *Institute for Creation Research*, 26 January 2009. http://www.icr.org/article/elephant-secrets-under-middle-east-sands.

Thomas, Herbert. *L'Homme avant l'Homme: le scénario des origines*. Paris: Gallimard, 1994.

——, Jack Roger, Sevket Sen, Chantal Bourdillon-De-Grissac, and Zaher Al-Sulaimani. 'Découverte de vertébrés fossiles dans l'Oligocène inférieur du Dhofar (Sultanat d'Oman)'. *Geobios* 22, No. 1 (1989): 101–20.

——, Jack Roger, Sevket Sen, and Zaher Al-Sulaimani. 'Découverte Des Plus Anciens "Anthropoides" Du Continent Arabo-africain Et D'un Primate

Tarsiiforme Dans l'Oligocène Du Sultanat d'Oman'. *Comptes rendus de l'Académie des sciences. Série II* 306 (1988): 823–9.

―――, Jack Roger, Sevket Sen, and Zaher Al-Sulaimani. 'Early Oligocene Vertebrates from Dhofar (Sultanate of Oman)'. In *Geology of the Arab World*, edited by Ali Sadek, 2: 283–92. Cairo: Cairo University, 1992.

―――, and Brigitte Senut. *Les Primates, ancêtres de l'Homme*. Paris: Artcom, 2001.

―――, Sevket Sen, Majeed Khan, Bernard Battail, and Giancarlo Ligabue. 'The Lower Miocene Fauna of As-Sarrar (Eastern Province, Saudi Arabia)'. *Atlal* 5 (1981): 109–36.

Thouless, Chris. 'Conservation in Saudi Arabia'. *Oryx* 25, n. 4 (1991): 222–8.

Tieleman, Irene. 'Avian adaptation along an aridity gradient: physiology, behavior, and life history'. PhD thesis, Rijksuniversiteit Groningen, 2002.

―――, and Joseph Williams. 'The Role of Hyperthermia in the Water Economy of Desert Birds'. *Physiological and Biochemical Zoology* 72, No. 1 (1999): 87–100.

Todorova, Vesela. 'Fossils help paint ancient portrait'. *The National*, 16 January 2010.

―――. '1997: Sheikh Zayed wins Gold Panda award'. *The National*, 6 November 2011. http://www.thenational.ae/news/uae-news/environment/1997-sheikh-zayed-wins-gold-panda-award.

―――. 'Emirati with a Passion to Keep UAE's Rare Bird Alive'. *The National*, 28 January 2013. http://www.thenational.ae/news/uae-news/environment/emirati-with-a-passion-to-keep-uaes-rare-bird-alive.

Al-Toraif, Mohammed. *Convention on the Conservation of Wildlife and Their Natural Habitats in the Countries of the Gulf Co-Operation Council*. Riyadh: Cooperation Council for the Arab States of the Gulf, 2009.

Turkish Journal of Botany. 'Scope of the Journal'. *Tübitak*, 2013. http://journals.tubitak.gov.tr/botany/scope.htm.

UAE Interact. 'Exhibition of ancient fossils opens in Abu Dhabi'. *UAE Interact*, 27 November 2005. http://www.uaeinteract.com/docs/Exhibition_of_ancient_fossils_opens_in_Abu_Dhabi/18743.htm.

Al-'Ubaydān, 'Abdallāh. 'Khādim al-ḥaramayn al-sharīfayn wa-sumūw walī al-'ahd tabarra'ā bi-15 milyūn lil-Hay'ah al-Waṭanīyah lil-Ḥayāt al-Fiṭrīyah'. *Al-Jazīrah*, 9 June 1998. http://www.al-jazirah.com/1998/19980609/p031.htm.

UNESCO. 'Falconry, a Living Human Heritage'. *UNESCO*, 2012. http://www.unesco.org/culture/ich/RL/00732.

―――. 'Arabian Oryx Sanctuary'. *UNESCO World Heritage Centre*, 2013. http://whc.unesco.org/en/list/654.

―――. 'World Heritage List'. *UNESCO World Heritage Centre*, 2013. http://whc.unesco.org/en/list.

United Nations Environment Programme. 'His Highness Sheikh Zayed Bin Sultan Al Nahyan'. *United Nations Environment Programme*, 2013. http://www.unep.org/champions/laureates/2005/Zayed.asp.

Vincent, Peter. *Saudi Arabia: An Environmental Overview*. London: Routledge, 2008.

Whybrow, Peter. 'Initiative for Funding Geological and Palaeontological Studies of Miocene Rocks, Baynunah Region, Western Province, Emirate of Abu Dhabi', January 1990. DF PAL/140/9/167. NHM Archives.

————. 'Brains in Abu Dhabi's Desert'. In *Travels with the Fossil Hunters*, edited by Peter Whybrow, 75–92. Cambridge: Cambridge University Press, 2000.

————, and Peter Andrews. Memorandum to H.W. Ball, 18 September 1981. DF PAL/140/9/167. NHM Archives.

————, and M.A. Bassiouni. 'The Arabian Miocene: rocks, fossils, primates and problems'. In *Primate Evolution*, edited by James Else and P.C. Lee, 85–91. Cambridge: Cambridge University Press, 1986.

————, and Andrew Hill. 'International Seminar on the Neogene vertebrates of Arabia' [1994]. DF PAL/140/9/167. NHM Archives.

————, and Andrew Hill, eds. *Fossil Vertebrates of Arabia: With Emphasis on the Late Miocene Faunas, Geology, and Palaeoenvironments of the Emirate of Abu Dhabi, United Arab Emirates*. New Haven: Yale University Press, 1999.

————, Andrew Hill, and Walid Yasin al Tikriti. 'Miocene Fossils From Abu Dhabi'. *Tribulus* 1, No. 1 (1991): 4–9.

————, Andrew Hill, Walid Yasin al-Tikriti, and Ernest A. Hailwood. 'Late Miocene primate fauna, flora and initial palaeomagnetic data from the Emirate of Abu Dhabi, United Arab Emirates'. *Journal of Human Evolution* 19 (1990): 582–8.

————, and H.A. McClure. 'Fossil mangrove roots and palaeoenvironments of the Miocene of the eastern Arabian Peninsula'. *Palaeogeography, Palaeoclimatology, Palaeoecology* 32 (1981): 213–25.

————, Andrew Smith, and Andrew Hill. 'The Fossil Record'. In *The Emirates: A Natural History*, edited by Peter Hellyer and Simon Aspinall, 80–9. Trident, 2005.

Williams, Joseph. 'A Phylogenetic Perspective of Evaporative Water Loss in Birds'. *The Auk* (1996): 457–72.

————. 'Curriculum Vitae'. Word document sent to me by the author, 11 June 2012.

————, and Irene Tieleman. 'Flexibility in Basal Metabolic Rate and Evaporative Water Loss Among Hoopoe Larks Exposed to Different Environmental Temperatures'. *Journal of Experimental Biology* 203, No. 20 (2000): 3153–9.

————. 'Physiological Ecology and Behavior of Desert Birds'. *Current Ornithology* 16 (2001): 299–353.

————. 'Physiological Adaptation in Desert Birds'. *BioScience* 55, No. 5 (2005): 416–25.

————, Irene Tieleman, and Mohammad Shobrak. 'Lizard Burrows Provide Thermal Refugia for Larks in the Arabian Desert'. *The Condor* 101 (1999): 714–17.

————, Mohammed Shobrak, Thomas Wilms, Ibrahim Arif, and Haseeb Khan. 'Climate change and animals in Saudi Arabia'. *Saudi Journal of Biological Sciences* 19, No. 2 (2012): 121–30.

Williamson, Douglas. 'The Relationship Between Taxonomy and Conservation'. In *Conservation of Arabian Gazelles*, edited by Arnaud Greth, Chris Magin, and Marc Ancrenaz, 45–50. Riyadh: NCWCD, 1996.

Winney, Bruce, Robert Hammond, William Macasero, Benito Flores, Ahmed Boug, Veronique Biquand, Sylvain Biquand, and Michael Bruford. 'Crossing the Red Sea: phylogeography of the hamadryas baboon, *Papio hamadryas hamadryas*'. *Molecular Ecology* 13, No. 9 (2004): 2819–27.

Wittmer, W. and W. Büttiker. 'Acknowledgements'. *Fauna of Saudi Arabia* 3 (1981), 3.

WWF. 'Emirates Wildlife Society in association with WWF (EWS-WWF)'. *WWF*, 2013. http://uae.panda.org/ews_wwf/index.cfm.
———. 'WWF International Directors'. *WWF*, 2013. http://wwf.panda.org/who_we_are/organization/directors/index.cfm.
Yapp, Carl. 'Adventurous oilman provides a role model for Bond'. *Western Mail*. Cardiff, 3 May 2002.
Zalmout, I.S., W.J. Sanders, L.M. MacLatchy, G.F. Gunnell, Y.A. Al-Mufarreh, M.A. Ali, A.A.H. Nasser, et al. 'New Oligocene primate from Saudi Arabia and the divergence of apes and Old World monkeys'. *Nature* 466, No. 7304 (2010): 360–4.
Zohary, Michael. 'On the Flora and Vegetation of the Middle East: Structure and Evolution'. In *Beiträge zur Umweltgeschichte des Vorderen Orients*, edited by Wolfgang Frey and Hans-Peter Uerpmann, 1–25. Wiesbaden: Reichert, 1981.
ZSL. 'Project Manager'. *New Scientist* 1619 (1988): 105.
———. 'Station Manager'. *New Scientist* 1627 (1988): 4.

## Secondary sources

Ahuja, Anjana. 'A New Golden Age Rises under the Desert Sun'. *The Telegraph*, 5 May 2012. http://www.telegraph.co.uk/science/9266426/A-new-golden-age-rises-under-the-desert-sun.html.
Akasoy, Anna. 'The Influence of the Arabic Tradition of Falconry and Hunting on Western Europe'. In *Islamic Medical and Scientific Tradition: Critical Concepts in Islamic Studies*, edited by Peter Pormann, 2: 406–27. Abingdon: Routledge, 2011.
El Alami, J., J.C. Dore, and J.F. Miquel. 'International Scientific Collaboration in Arab Countries'. *Scientometrics* 23, No. 1 (1992): 249–63.
Altbach, Philip. 'Centers and Peripheries in the Academic Profession: The Special Challenges of Developing Countries'. In *The Decline of the Guru: The Academic Profession in the Third World*, 1–21. Edited by Philip Altbach. New York: Palgrave Macmillan, 2003.
Andersen, Casper, Jakob Bek-Thomsen, and Peter Kjærgaard. 'The Money Trail: A New Historiography for Networks, Patronage, and Scientific Careers'. *Isis* 103, No. 2 (2012): 310–15.
Anscombe, Frederick. *The Ottoman Gulf: The Creation of Kuwait, Saudi Arabia, and Qatar*. New York: Columbia University Press, 1993.
Asghar, Anila, Jason Wiles, and Brian Alters. 'The origin and evolution of life in Pakistani High School Biology'. *Journal of Biological Education* 44, No. 2 (2010): 65–71.
Atar, Mohammad. 'Quest for Identity: The Role of the Textbook in Forming Saudi Arabian Identity'. PhD dissertation, University of Oregon, 1988.
Al-Atawneh, Muhammad. *Waḥḥābī Islam Facing the Challenges of Modernity: Dār Al-Iftā in the Modern Saudi State*. Leiden: Brill, 2010.
Atwan, Abdul Bari. 'In the realm of the censor'. *Index on Censorship* 25, No. 4 (1996): 77–9.
Ayoub, Mahmoud. 'Creation or Evolution? The Reception of Darwinism in Modern Arab Thought'. In *Science and Religion in a Post-colonial World: Interfaith Perspectives*, edited by Zainal Abidin Bagir, 173–90. Hindmarsh: ATF, 2005.
Al-Azm, Sadik. 'Islam and the science–religion debates in modern times'. *European Review* 15, No. 3 (2007): 283–95.

Babar, Zahra. 'The Cost of Belonging: Citizenship Construction in the State of Qatar'. *Middle East Journal* 68, No. 3 (2014): 403–20.

Badran, Adnan. 'The Arab States'. In *UNESCO Science Report 2005*, edited by Mustafa El Tayeb, 159–176. Paris: Unesco, 2005.

Ballantyne, Tony. 'Race and the Webs of Empire: Aryanism from India to the Pacific'. *Journal of Colonialism and Colonial History* 2, No. 3 (2001).

Barabási, A.L., H. Jeong, Z. Néda, E. Ravasz, A. Schubert, and T. Vicsek. 'Evolution of the social network of scientific collaborations'. *Physica A* 311, No. 3–4 (2002): 590–614.

Barton, Gregory. 'Environmentalism, Development and British Policy in the Middle East 1945–65'. *Journal of Imperial and Commonwealth History* 38, No. 4 (2010): 619–39.

Basalla, George. 'The Spread of Western Science: A three-stage model describes the introduction of modern science into any non-European nation'. *Science* 156, No. 3775 (1967): 611–22.

Āl Bassām, ʿAbd Allāh ibn ʿAbd al-Raḥmān ibn Ṣāliḥ, ed. *Khizānat al-tawārīkh al-Najdīyah*. Vol. 7. 1999.

Bates, Stephen. 'Prince Charles still enjoy thrill of the chase while still legal despite mother's advice'. *The Guardian*, 6 November 2004, http://www.guardian.co.uk/uk/2004/nov/06/monarchy.hunting.

Beaudevin, Claire, 'Old diseases & contemporary crisis. Inherited blood disorders in the Sultanate of Oman'. *Anthropology & Medicine* 20, No. 3 (2013).

Beblawi, Hazem. 'The Rentier State in the Arab World'. *Arab Studies Quarterly* 9, No. 4 (1987): 383–98.

———, and Giacomo Luciani, eds. *The Rentier State*. London: Croom Helm, 1987.

Bennett, Brett, and Joseph Hodge, eds. *Science and Empire: Knowledge and Networks of Science across the British Empire, 1800–1970*. Basingstoke: Palgrave Macmillan, 2011.

Bismarck, Helene von. *British Policy in the Persian Gulf, 1961–1968: Conceptions of Informal Empire*. Basingstoke: Palgrave Macmillan, 2013.

Bizri, Omar. 'Research, Innovation, Entrepreneurship and the Rentier Culture in the Arab Countries'. In *The Real Issues of the Middle East and the Arab Spring*, edited by Thomas Andersson and Abdelkader Djeflat, 195–227. New York: Springer, 2013.

Bond, Michael. 'Where progress is a lost cause'. *New Scientist* 178, No. 2392 (2003): 27.

BouJaoude, Saouma, Anila Asghar, Jason Wiles, Lama Jaber, Diana Sarieddine, and Brian Alters. 'Biology Professors' and Teachers' Positions Regarding Biological Evolution and Evolution Education in a Middle Eastern Society'. *International Journal of Science Education* 33, No. 7 (2011): 979–1000.

———, Jason Wiles, Anila Asghar, and Brian Alters. 'Muslim Egyptian and Lebanese Students' Conceptions of Biological Evolution'. *Science & Education* 20, No. 9 (2011): 895–915.

Bourrelier, Paul-Henri, and Jean Lespine. 'Les opérations minières outre-mer: Le BRGM, acteur central de la politique publique'. *Réalités Industrielles* (August 2008): 7–20.

Buj, Antonio. 'International experimentation and control of the locust plague: Africa in the first half of the 20th century'. In *Nature et environnement*, edited by Yvon Chatelin and Christophe Bonneuil, 3: 93–105. Paris: Orstom, 1995.

Burton, Elise. 'Teaching Evolution in Muslim States: Iran and Saudi Arabia Compared'. *Reports of the National Center for Science Education* 30, No. 3 (2010): 25–9.

———. 'Evolution and Creationism in Middle Eastern Education: A New Perspective'. *Evolution* 65, No. 1 (2010): 1–4.

———. 'Science, Religion and the State: Teaching Evolution in the Middle East'. Senior Honors Thesis, University of California, Berkeley, 2010.

Butler, Declan. 'Can a Saudi university think freely?' *Nature* 447, No. 7146 (2007): 758–9.

Callon, M., and V. Rabeharisoa. 'Research "in the Wild" and the Shaping of New Social Identities'. *Technology in Society* 25, No. 2 (2003): 193–204.

Cittadino, Eugene. *Nature as the laboratory: Darwinian plant ecology in the German Empire, 1880–1900.* Cambridge: Cambridge University Press, 1990.

Clark, Janine. 'Social Movement Theory and Patron-Clientelism: Islamic Social Institutions and the Middle Class in Egypt, Jordan, and Yemen'. *Comparative Political Studies* 37, No. 8 (2004): 941–68.

Commins, David. *The Gulf States: A Modern History.* London: I.B.Tauris, 2012.

Cook, Michael. 'Ibn Qutayba and the monkeys'. *Studia Islamica* No. 89 (1999): 43–74.

Dagher, Zoubeida, and Saouma BouJaoude. 'Scientific Views and Religious Beliefs of College Students: The Case of Biological Evolution'. *Journal of Research in Science Teaching* 34, No. 5 (1997): 429–45.

Dallal, Ahmad. *Islam, Science, and the Challenge of History.* New Haven: Yale University Press, 2010.

Davidson, Christopher. *Abu Dhabi: Oil and Beyond.* London: Hurst, 2009.

———. *The Persian Gulf and Pacific Asia: From Indifference to Interdependence.* New York: Columbia University Press, 2010.

———. *Persian Gulf–Pacific Asia Linkages in the Twenty-First Century: A Marriage of Convenience?* London: London School of Economics and Political Science, 2010.

———. *After the Sheikhs: The Coming Collapse of the Gulf Monarchies.* London: Hurst, 2012.

Davis, Diana. 'Environmentalism as Social Control? An Exploration of the Transformation of Pastoral Nomadic Societies in French Colonial North Africa'. *The Arab World Geographer* 3, No. 3 (2000): 182–98.

———. 'Indigenous knowledge and the desertification debate: problematising expert knowledge in North Africa'. *Geoforum* 36, No. 4 (2005): 509–24.

———. *Resurrecting the Granary of Rome: Environmental History and French Colonial Expansion in North Africa.* Athens: Ohio University Press, 2007.

———. 'Brutes, beasts and empire: veterinary medicine and environmental policy in French North Africa and British India'. *Journal of Historical Geography* 34, No. 2 (2008): 242–67.

———. 'Enclosing Nature in North Africa: National Parks and the Politics of Environmental History'. In *Water on Sand: Environmental Histories of the Middle East and North Africa,* edited by Alan Mikhail, 159–79. New York: Oxford University Press, 2013.

Dean, Cornelia. 'Islamic Creationist and a Book Sent Round the World'. *The New York Times,* 17 July 2007. http://www.nytimes.com/2007/07/17/science/17book.html.

Degenne, Alain, and Michel Forsé. *Introducing Social Networks.* London: Sage, 1999.

Desmond, Ray. *The History of the Royal Botanic Gardens Kew.* Second edition. Kew, 2007.

Dobzhansky, Theodosius. 'Nothing in Biology Makes Sense Except in the Light of Evolution'. *American Biology Teacher* 35, No. 3 (1973): 125–9.

Dolbee, Samuel, and Chris Gratien. 'Nation, Class, and Ecology in French Mandate Lebanon: AUB and 1930s Rural Development'. *Ottoman History Podcast*, No. 59 (2012). http://www.ottomanhistorypodcast.com/2012/07/nation-class-and-ecology-in-french_07.html.

Donn, Gari, and Yahya Al Manthri. *Globalization and Higher Education in the Arab Gulf States.* Didcot: Symposium, 2010.

Drayton, Richard. *Nature's Government: Science, Imperial Britain, and the 'Improvement' of the World.* New Haven: Yale University Press, 2000.

Dresch, Paul, and James Piscatori, eds. *Monarchies and Nations: Globalization and Identity in the Arab States of the Gulf.* London: I.B.Tauris, 2005.

Dundar, Fuat. 'Empire of Taxonomy: Ethnic and Religious Identities in the Ottoman Surveys and Censuses'. *Middle Eastern Studies* (2014). doi:10.1080/002 63206.2014.913134.

Edis, Taner. 'Cloning Creationism in Turkey'. *Reports of the National Center for Science Education* 19, No. 6 (1999): 30–5.

———. *An Illusion of Harmony: Science and Religion in Islam.* Amherst: Prometheus, 2007.

———. 'Modern Science and Conservative Islam: An Uneasy Relationship'. *Science & Education* 18, No. 6–7 (2009): 885–903.

Eisenstein, Herbert. 'Die Trappe in der klassisch-arabischen Literatur'. *Wiener Zeitschrift für die Kunde des Morgenlandes* 75 (1983): 35–64.

———. *Einführung in die arabische Zoographie: Das tierkundliche Wissen in der arabisch-islamischen Literatur.* Berlin: Dietrich Reimer, 1991.

———. 'Some Accounts of Zoological Experiments in Classical Arabic Literature'. *Arabist: Budapest Studies in Arabic* 15–16 (1995): 113–20.

———. 'Die arabische Tierkunde – Zoologie oder Zoographie?' *Wiener Zeitschrift für die Kunde des Morgenlandes* 97 (2007): 147–61.

Elshakry, Marwa. 'The Gospel of Science and American Evangelism in Late Ottoman Beirut'. *Past and Present* 196, No. 1 (2007): 173–214.

———. 'The Exegesis of Science in Twentieth-Century Arabic Interpretations of the Qurʾān'. In *Nature and Scripture in the Abrahamic Religions: 1700–Present*, edited by Jitse van der Meer and Scott Mandelbrote, 2: 491–523. Leiden: Brill, 2008.

———. 'Global Darwin: Eastern enchantment'. *Nature* 461, No. 7268 (2009): 1200–1.

———. 'When Science Became Western: Historiographical Reflections'. *Isis* 101 (2010): 98–109.

———. 'Darwinian Conversions: Science and Translation in Late Ottoman Egypt and Greater Syria'. In *Perilous Modernity: History of Medicine in the Ottoman Empire and the Middle East from the 19th Century Onwards*, edited by Anne Moulin and Yeşim Ulman, 85–95. Istanbul: Isis, 2010.

———. 'Muslim Hermeneutics and Arabic Views of Evolution'. *Zygon* 46, No. 2 (2011): 330–44.

———. 'Early Arabic Views of Darwin'. In *Science and Religion: Christian and Muslim Perspectives*, edited by David Marshall, 128–33. Washington: Georgetown University Press, 2012.

————. *Reading Darwin in Arabic, 1860–1950*. Chicago: University of Chicago Press, 2013.

Ende, Werner. 'Religion, Politik und Literatur in Saudi-Arabien: Der geistes-geschichtliche Hintergrund der heutigen religiösen und kulturpolitischen Situation (III)'. *Orient* 23, No. 3 (1982): 378–93.

Ewers, Michael, and Edward Malecki. 'Leapfrogging into the Knowledge Economy: Assessing the Economic Development Strategies of the Arab Gulf States'. *Tijdschrift voor Economische en Sociale Geografie* 101, No. 5 (2010): 494–508.

Faheem, Mohammed. 'Higher Education and Nation Building: A Case Study of King Abdulaziz University'. PhD thesis, University of Illinois at Urbana-Champaign, 1982.

Fan, Fa-ti. 'The Global Turn in the History of Science'. *East Asian Science, Technology and Society* 6, No. 2 (2012): 249–58.

Fergany, Nader. 'Science and Research for Development in the Arab Region'. In *Research for Development in the Middle East and North Africa*, edited by Eglal Rached and Dina Craissati, 61–110. Ottawa: International Development Research Centre, 2000.

Findlow, Sally. 'International networking in the United Arab Emirates higher education system: Global–local tensions'. *Compare* 35, No. 3 (2005): 285–302.

Foley, Sean. *The Arab Gulf States: Beyond Oil and Islam*. Boulder: Lynne Rienner, 2010.

Foltz, Richard. 'Islamic Environmentalism: A Matter of Interpretation'. In *Islam and Ecology: A Bestowed Trust*, edited by Richard Foltz, Frederick Denny, and Azizan Baharuddin, 249–79. Cambridge: Harvard University Press, 2003.

————. 'Introduction: The Environmental Crisis in the Muslim World'. In *Environmentalism in the Muslim World*, edited by Richard Foltz, vii–xiii. New York: Nova Science, 2005.

Frankel, O.H., and Michael E. Soulé. *Conservation and Evolution*. Cambridge: Cambridge University Press, 1981.

Fromherz, Allen. *Qatar: A Modern History*. London: I.B.Tauris, 2012.

Gari, Lutfallah. 'A History of the Himā Conservation System'. *Environment and History* 12, No. 2 (2006): 213–28.

Gause, Gregory. 'Saudi Arabia over a Barrel'. *Foreign Affairs* 79, No. 3 (2000): 80–94.

Ghabra, Shafeeq, and Margreet Arnold. *Studying the American Way: An Assessment of American-Style Higher Education in Arab Countries*. Policy Focus 71. Washington: Washington Institute for Near East Policy, 2007.

Glaisyer, Natasha. 'Networking: trade and exchange in the eighteenth-century British empire'. *Historical Journal* 47, No. 2 (2004): 451–76.

Golshani, Mehdi. 'Islam and the Sciences of Nature: Some Fundamental Questions'. *Islamic Studies* 39, No. 4 (2000): 597–611.

Granovetter, Mark. 'The Strength of Weak Ties'. *American Journal of Sociology* 78, No. 6 (1973): 1360–80.

Gray, Matthew. *A Theory of 'Late Rentierism' in the Arab States of the Gulf*. Occasional Paper 7. Center for International and Regional Studies, Georgetown University School of Foreign Service in Qatar, 2011.

Grove, Richard. *Green Imperialism: Colonial Expansion, Tropical Island Edens and the Origins of Environmentalism, 1600–1860*. Cambridge: Cambridge University Press, 1995.

Grutz, Jane. 'A King and Two Salukis'. *Saudi Aramco World* 59, No. 3 (2008): 40–7.

Guessoum, Nidhal. 'The Qur'an, Science, and the (related) Contemporary Muslim Discourse'. *Zygon* 43, No. 2 (2008): 411–31.

———. *Islam's Quantum Question: Reconciling Muslim Tradition and Modern Science.* London: I.B.Tauris, 2011.

Hameed, Salman. 'Evolution and creationism in the Islamic world'. In *Science and Religion: New Historical Perspectives*, edited by Thomas Dixon, Geoffrey Cantor, and Stephen Pumfrey, 133–52. Cambridge: Cambridge University Press, 2010.

Hanafi, Sari. 'University systems in the Arab East: Publish globally and perish locally vs publish locally and perish globally'. *Current Sociology* 59, No. 3 (2011): 291–309.

Harbinson, David. 'The US – Saudi Arabian Joint Commission on Economic Cooperation: A Critical Appraisal'. *Middle East Journal* 44, No. 2 (1990): 269–83.

Harris, Steven. 'Long-Distance Corporations, Big Sciences, and the Geography of Knowledge'. *Configurations* 6, No. 2 (1998): 269–304.

Hayhurst, John. 'A Quest for Knowledge: The Basra Date Palm, the Botanical Garden in Bengal'. *Qatar Digital Library*, 14 October 2014. http://www.qdl.qa/en/quest-knowledge-basra-date-palm-botanical-garden-bengal.

Helmy Mohammad, A.H. 'Notes on the Reception of Darwinism in Some Islamic Countries'. In *Science in Islamic Civilization*, edited by Ekmeleddin İhsanoğlu and Feza Günergun, 245–55. İstanbul: Research Centre for Islamic History, Art and Culture, 2000.

Herb, Michael. *All in the Family: Absolutism, Revolution, and Democracy in the Middle Eastern Monarchies*. Albany: State University of New York Press, 1999.

Hertog, Steffen. 'Shaping the Saudi State: Human Agency's Shifting Role in Rentier-State Foundation'. *International Journal of Middle East Studies* 39, No. 4 (2007): 539–63.

———. *Princes, Brokers, and Bureaucrats: Oil and the State in Saudi Arabia.* Ithaca: Cornell University Press, 2010.

———. 'The Sociology of the Gulf Rentier Systems: Societies of Intermediaries'. *Comparative Studies in Society and History* 52, No. 2 (2010): 282–318.

———. 'Defying the Resource Curse: Explaining Successful State-Owned Enterprises in Rentier-States'. *World Politics* 62, No. 2 (2010): 261–301.

———. 'Lean and mean: The new breed of state-owned enterprises in the Gulf monarchies'. In *Industrialization in the Gulf: A Socioeconomic Revolution*, edited by Jean-François Seznec and Mimi Kirk, 17–29. Abingdon: Routledge, 2011.

———. 'How the GCC did it: formal and informal governance of successful public enterprise in the Gulf Co-operation Council countries'. In *Towards New Arrangements for State Ownership in the Middle East and North Africa*, 71–92. OECD Publishing, 2012.

———. 'State and Private Sector in the GCC after the Arab Uprisings'. *Journal of Arabian Studies* 3, No. 2 (2013): 174–95.

———. *Arab Gulf States: An Assessment of Nationalisation Policies.* Gulf Labour Markets and Migration Programme, 2014.

Hiepko, Paul. 'Die Sammlungen des Botanischen Museums Berlin-Dahlem und ihre Geschichte'. In *Geschichte der Botanik in Berlin*, edited by Claus Schnarrenberger and Hildemar Scholz, 297–318. Berlin: Colloquium, 1990.

Hirschler, Konrad. *Medieval Arabic Historiography: Authors as actors*. Abingdon: Routledge, 2006.

Hodge, Joseph. 'Science, Development, and Empire: The Colonial Advisory Council on Agriculture and Animal Health, 1929–43'. *Journal of Imperial and Commonwealth History* 30, No. 1 (2002): 1–26.

———. 'British Colonial Expertise, Post-Colonial Careering and the Early History of International Development'. *Journal of Modern European History* 8, No. 1 (2010): 24–46.

Hoffman, Andrew. *From Heresy to Dogma: An Institutional History of Corporate Environmentalism*. Stanford: Stanford University Press, 2001.

Howard, Damian. *Being Human in Islam: The Impact of the Evolutionary Worldview.* Abingdon: Routledge, 2011.

Joma, Hesam Addin. 'The Earth as a mosque: Integration of the traditional Islamic environmental planning ethic with agricultural and water development policies in Saudi Arabia'. PhD dissertation, University of Pennsylvania, 1991.

Jones, Andrew. *Developmental Fairy Tales: Evolutionary Thinking and Modern Chinese Culture.* Cambridge: Harvard University Press, 2011.

Jones, Toby. *Desert Kingdom: How Oil and Water Forged Modern Saudi Arabia.* Cambridge: Harvard University Press, 2010.

Kadushin, Charles. *Understanding Social Networks: Theories, Concepts, and Findings.* Oxford: Oxford University Press, 2011.

Kamrava, Mehran. *Qatar: Small State, Big Politics.* Ithaca: Cornell University Press, 2013.

Kéchichian, Joseph. *Legal and Political Reforms in Sa'udi Arabia.* Abingdon: Routledge, 2013.

Kjærgaard, Peter. '"Hurrah for the Missing Link!": A History of Apes, Ancestors and a Crucial Piece of Evidence'. *Notes and Records of the Royal Society* 65, No. 1 (2011): 83–98.

———. 'The Fossil Trade: Paying a Price for Human Origins'. *Isis* 103, No. 2 (2012): 340–55.

King, David. *Islamic Astronomy and Geography.* Farnham: Ashgate, 2012.

Klaver, J.M.I. *Scientific Expeditions to the Arab World (1761–1881).* London: Arcadian Library, 2009.

Kohler, Robert. 'Place and Practice in Field Biology'. *History of Science* 40, No. 2 (2002): 189–210.

Krawietz, Birgit. 'Falconry as Cultural Icon of the Arab Gulf Region'. Manuscript sent to me by the author in April 2013.

Kritsky, Gene. 'Teaching Evolution in an Islamic Country'. *The American Biology Teacher* 46, No. 5 (1984): 266–71.

Kruk, Remke. 'Traditional Islamic Views of Apes and Monkeys'. In *Ape, Man, Apeman: Changing Views Since 1600*, edited by Raymond Corbey and Bert Theunissen, 29–41. Leiden: Leiden University, 1995.

———. 'Ibn Abī l-Ash'ath's *Kitāb al-Ḥayawān*: A Scientific Approach to Anthropology, Dietetics and Zoological Systematics'. *Zeitschrift für Geschichte der Arabisch-Islamischen Wissenschaften* 14 (2001): 119–68.

Lacey, Robert. *Inside the Kingdom: Kings, Clerics, Modernists, Terrorists and the Struggle for Saudi Arabia.* London: Hutchinson, 2009.

Lacroix, Stéphane. *Les islamistes saoudiens: Une insurrection manquée.* Paris: Presses universitaires de France, 2010.

Latour, BruNo. *Science in Action: How to follow scientists and engineers through society.* Cambridge: Harvard University Press, 1987.

————. *Pandora's Hope: Essays on the Reality of Science Studies*. Cambridge: Harvard University Press, 1999.

Laursen, Lucas. 'Fossil skull fingered as ape–monkey ancestor'. *Nature News*, 14 July 2010. doi:10.1038/news.2010.354.

Lawler, Andrew. 'In search of Green Arabia'. *Science* 345, No. 6200 (2014): 994–7.

Lester, Alan. 'Imperial Circuits and Networks: Geographies of the British Empire'. *History Compass* 4, No. 1 (2006): 124–41.

Leydesdorff, Loet, and Caroline Wagner. 'International collaboration in science and the formation of a core group'. *Journal of Informetrics* 2, No. 4 (2008): 317–25.

Library of Congress. 'ALA-LC Romanization Tables'. *The Library of Congress*, 2010. http://www.loc.gov/catdir/cpso/roman.html.

Lippman, Thomas. 'Cooperation under the Radar: The US–Saudi Arabian Joint Commission for Economic Cooperation'. *Middle East Institute*, 1 October 2009. http://www.mei.edu/content/cooperation-under-radar-us-saudi-arabian-joint-commission-economic-cooperation-jecor.

Lipps, Jere. 'What, if anything, is micropaleontology?' *Paleobiology* 7, No. 2 (1981): 167–99.

Livingston, John. 'Western Science and Educational Reform in the Thought of Shaykh Rifaʻa al-Tahtawi'. *International Journal of Middle East Studies* 28, No. 4 (1996): 543–64.

Livingstone, David. *Putting Science in Its Place: Geographies of Scientific Knowledge*. Chicago: University of Chicago Press, 2003.

Llewellyn, Othman. 'The Basis for a Discipline of Islamic Environmental Law'. In *Islam and Ecology: A Bestowed Trust*, edited by Richard Foltz, Frederick Denny, and Azizan Baharuddin, 185–247. Cambridge: Harvard University Press, 2003.

Lotfalian, Mazyar. *Islam, Technoscientific Identites, and the Culture of Curiosity*. Lanham: University Press of America, 2004.

Lowe, Daniel. 'The King's Oryx: Ibn Saud's Diplomatic Gift to George V'. *Qatar Digital Library*, 14 October 2014. http://www.qdl.qa/en/king%E2%80%99s-oryx-ibn-saud%E2%80%99s-diplomatic-gift-george-v.

Luomi, Mari. *The Gulf Monarchies and Climate Change: Abu Dhabi and Qatar in an Era of Natural Unsustainability*. London: Hurst, 2012.

Lux, David, and Harold Cook. 'Closed Circles or Open Networks?: Communicating at a Distance During the Scientific Revolution'. *History of Science* 36, No. 2 (1998): 179–211.

Matthiesen, Toby. *Sectarian Gulf: Bahrain, Saudi Arabia, and the Arab Spring That Wasn't*. Stanford: Stanford University Press, 2013.

Mazawi, André. 'Divisions of Academic Labor: National and Non-national Faculty Members in Arab Gulf Universities'. *International Journal of Contemporary Sociology* 40, No. 1 (2003): 91–110.

————. 'The Academic Workplace in Public Arab Gulf Universities'. In *The Decline of the Guru: The Academic Profession in the Third World*, edited by Philip Altbach, 231–69. New York: Palgrave Macmillan, 2003.

McDonald, M.V. 'Animal-Books as a Genre in Arabic Literature'. *British Society for Middle Eastern Studies Bulletin* 15 (1988): 3–10.

McNeill, J.R., and William McNeill. *The Human Web: A Bird's-Eye View of World History*. New York: W.W. Norton, 2003.

Mellahi, Kamel, and Said Al-Hinai. 'Local Workers in Gulf Co-operation Countries: Assets or Liabilities?' *Middle Eastern Studies* 36, No. 3 (2000): 177–90.

Ménoret, Pascal. *Joyriding in Riyadh: Oil, Urbanism, and Road Revolt.* New York: Cambridge University Press, 2014.

Mikhail, Alan. *The Animal in Ottoman Egypt.* New York: Oxford University Press, 2014.

Milgram, Stanley. 'The Small-World Problem'. *Psychology Today* 2, No. 1 (1967): 60–7.

Al-Misnad, Sheikha. *The Development of Modern Education in the Gulf.* London: Ithaca, 1985.

Mitchell, Timothy. *Colonising Egypt.* Berkeley: University of California Press, 1991.

———. *Rule of Experts: Egypt, Techno-Politics, Modernity.* Berkeley: University of California Press, 2002.

———. *Carbon Democracy: Political Power in the Age of Oil.* London: Verso, 2013.

Nazim Ali, S. 'Acquisition of Scientific Literature in Developing Countries: 5: Arab Gulf Countries'. *Information Development* 5, No. 2 (1989): 108–15.

Newman, M.E. 'The structure of scientific collaboration networks'. *Proceedings of the National Academy of Sciences of the United States of America* 98, No. 2 (2001): 404–9.

Ninham, Sally. *A Cohort of Pioneers: Australian postgraduate students and American postgraduate degrees 1949–1964.* Ballan: Connor Court, 2012.

Nyhart, Lynn. *Modern Nature: The Rise of the Biological Perspective in Germany.* Chicago: University of Chicago Press, 2009.

Ochsenwald, William. 'Saudi Arabia and the Islamic Revival'. *International Journal of Middle East Studies* 13, No. 3 (1981): 271–86.

Al-Omar, Abdullah. 'The Reception of Darwinism in the Arab World'. PhD dissertation, Harvard University, 1982.

Onley, James. *The Arabian Frontier of the British Raj: Merchants, Rulers, and the British in the Nineteenth Century Gulf.* Oxford: Oxford University Press, 2007.

Orth, Karin. 'Das Förderprofil der Deutschen Forschungsgemeinschaft 1949 bis 1969'. *Berichte zur Wissenschaftsgeschichte* 27, No. 4 (2004): 261–83.

Ouis, Pernilla. '"Greening the Emirates": the modern construction of nature in the United Arab Emirates'. *Cultural Geographies* 9, No. 3 (2002): 334–47.

Parker, Chad. 'Controlling man-made malaria: Corporate modernisation and the Arabian American Oil Company's malaria control program in Saudi Arabia, 1947–1956'. *Cold War History* 12, No. 3 (2012): 473–94.

Peker, Deniz, Gulsum Gul Comert, and Aykut Kence. 'Three Decades of Anti-evolution Campaign and its Results: Turkish Undergraduates' Acceptance and Understanding of the Biological Evolution Theory'. *Science & Education* 19, No. 6–8 (2010): 739–55.

Potter, Simon. 'Webs, Networks, and Systems: Globalization and the Mass Media in the Nineteenth- and Twentieth-Century British Empire'. *Journal of British Studies* 46, No. 3 (2007): 621–46.

Prendergast, David, and William Adams. 'Colonial Wildlife Conservation and the Origins of the Society for the Preservation of the Wild Fauna of the Empire (1903–1914)'. *Oryx* 37, No. 2 (2003): 251–60.

Prokop, Michaela. 'Political Economy of Fiscal Crisis in a Rentier State: Case Study of Saudi Arabia'. PhD thesis, University of Durham, 1999.

Raj, Kapil. *Relocating Modern Science: Circulation and the Construction of Knowledge in South Asia and Europe, 1650–1900.* Basingstoke: Palgrave Macmillan, 2007.

——. 'Beyond Postcolonialism ... and Postpositivism: Circulation and the Global History of Science'. *Isis* 104, No. 2 (2013): 337–47.

Al-Rasheed, Madawi, ed. *Transnational Connections and the Arab Gulf*. Abingdon: Routledge, 2005.

——. 'Transnational Connections and National Identity: Zanzibari Omanis in Muscat'. In *Monarchies and Nations: Globalization and Identity in the Arab States of the Gulf*, edited by Paul Dresch and James Piscatori, 96–113. London: I.B.Tauris, 2005.

——. *Kingdom without Borders: Saudi political, religious and media frontiers*. London: Hurst, 2008.

——. *A Most Masculine State: Gender, Politics and Religion in Saudi Arabia*. Cambridge: Cambridge University Press, 2013.

Le Renard, Amélie. '"Only for Women:" Women, the State, and Reform in Saudi Arabia'. *Middle East Journal* 62, No. 4 (2008): 610–29.

Riexinger, Martin. 'Propagating Islamic Creationism on the Internet'. *Masaryk University Journal of Law and Technology* 2, No. 2 (2008): 99–112.

——. 'Responses of South Asian Muslims to the Theory of Evolution'. *Die Welt des Islams* 49, No. 2 (2009): 212–47.

——. 'Islamic Opposition to the Darwinian Theory of Evolution'. In *Handbook of Religion and the Authority of Science*, edited by Olav Hammer and Jim Lewis, 483–510. Leiden: Brill, 2010.

Roberts, Lissa. 'Situating Science in Global History: Local Exchanges and Networks of Circulation'. *Itinerario* 33, No. 1 (2009): 9–30.

Röllig, Wolfgang. 'Der Tübinger Atlas des Vorderen Orients und seine altorientalischen Karten'. *Acta Antiqua Academiae Scientiarum Hungaricae* 22 (1974): 531–7.

Roman, Howaida. *Emigration Policy in Egypt*. European University Institute, 2006.

Sabra, A.I. 'Situating Arabic Science: Locality versus Essence'. *Isis* 87, No. 4 (1996): 654–70.

Said, Edward. *Orientalism*. London: Routledge, 1978.

Sardar, Ziauddin. 'Middle East brain drain switches back from the West'. *Nature* 283, No. 5745 (1980): 327–28.

Satia, Priya. '"A Rebellion of Technology": Development, Policing, and the British Arabian Imaginary'. In *Environmental Imaginaries of the Middle East and North Africa*, edited by Diana K. Davis and Edmund Burke, 23–59. Athens: Ohio University Press, 2011.

Schayegh, Cyrus. *Who Is Knowledgeable, Is Strong: Science, Class, and the Formation of Modern Iranian Society, 1900–1950*. Berkeley: University of California Press, 2009.

Scott, James. *Seeing Like a State: How Certain Schemes to Improve the Human Condition Have Failed*. New Haven: Yale University Press, 1998.

Sepkoski, David. *Rereading the Fossil Record: The Growth of Paleobiology as an Evolutionary Discipline*. Chicago: University of Chicago Press, 2012.

Sezgin, Fuat. *Wissenschaft und Technik im Islam*. Vol. 1. Frankfurt: Institut für Geschichte der Arabisch-Islamischen Wissenschaften, 2003.

El Shakry, Omnia. *The Great Social Laboratory: Subjects of Knowledge in Colonial and Postcolonial Egypt*. Stanford: Stanford University Press, 2007.

El-Shibiny, Mohamed. 'Higher Education in Oman: Its Development and Prospects'. In *Higher Education in the Gulf: Problems and Prospects*, edited by K.E. Shaw, 150–81. Exeter: University of Exeter Press, 1997.

Sivasundaram, Sujit. *Nature and the Godly Empire: Science and Evangelical Mission in the Pacific, 1795–1850*. Cambridge: Cambridge University Press, 2005.

Skovgaard-Petersen, Jakob, and Bettina Gräf, eds. *The Global Mufti: The Phenomenon of Yusuf al-Qaradawi*. New York: Columbia University Press, 2009.

Sörlin, Sverker. 'National and International Aspects of Cross-Boundary Science: Scientific Travel in the 18th Century'. In *Denationalizing Science: The Contexts of International Scientific Practice*, edited by Elisabeth Crawford, Terry Shinn, and Sverker Sörlin, 43–72. Dordrecht: Kluwer, 1993.

Springborg, Robert. 'GCC Countries as "Rentier States" Revisited'. *Middle East Journal* 67, n. 2 (2013): 301–9.

Stenberg, Leif. *The Islamization of Science: Four Muslim Positions Developing an Islamic Modernity*. Stockholm: Almqvist & Wiksell International, 1996.

Stevenson, Angus, ed. *Oxford Dictionary of English*. Oxford: Oxford University Press, 2010.

———, ed. 'Social network'. *Oxford Dictionary of English*. Oxford: Oxford University Press, 2010. http://www.oxfordreference.com/10.1093/acref/9780199571123.001.0001/m_en_gb0994346.

Stolz, Daniel. '"By Virtue of Your Knowledge": Scientific Materialism and the Fatwās of Rashīd Riḍā'. *Bulletin of the School of Oriental and African Studies* 75, No. 2 (2012): 223–47.

Stremmel, Fabian. 'An Imperial German Battle to Win over Mesopotamia: The Baghdad *Propagandaschule* (1909–17)'. *Middle Eastern Studies* (2014).

Tamraz, Ahmad. 'A Study of Availability and Actual Usage of Arabic and English Monographs in Science and Technology in Three Academic Libraries in Saudi Arabia'. PhD dissertation, Rutgers, 1984.

Teitelbaum, Joshua. 'Dueling for "Da'wa": State vs. Society on the Saudi Internet'. *Middle East Journal* 56, No. 2 (2002): 222–39.

Tétreault, Mary Ann. 'Identity and Transplant-University Education in the Gulf: The American University of Kuwait'. *Journal of Arabian Studies* 1, No. 1 (2011): 81–98.

Timler, Friedrich Karl, and Bernhard Zepernick. 'German Colonial Botany'. In *Beiträge zur neueren Geschichte der Botanik in Deutschland*, edited by H. Lorenzen, 143–68. Stuttgart: Gustav Fischer, 1987.

Torstrick, Rebecca, and Elizabeth Faier. *Culture and Customs of the Arab Gulf States*. Westport: Greenwood, 2009.

Turchetti, Simone, Néstor Herran, and Soraya Boudia. 'Introduction: have we ever been 'transnational'? Towards a history of science across and beyond borders'. *British Journal for the History of Science* 45, No. 3 (2012): 319–36.

United Nations Development Programme. *The Arab Human Development Report 2003: Building a Knowledge Society*. New York: United Nations Development Programme, 2003.

University of Chicago Press. *The Chicago Manual of Style*. Sixteenth ed. Chicago: University of Chicago Press, 2010.

Viré, F. 'Ḳird'. In *Encyclopaedia of Islam*, edited by C.E. Bosworth, E. van Donzel, B. Lewis, and Ch. Pellat, New Edition, 5: 131–4. Leiden: Brill, 1986.

Vitalis, Robert, and Steven Heydemann. 'War, Keynesianism, and Colonialism: Explaining State-Market Relations in the Postwar Middle East'. In *War, Institutions, and Social Change in the Middle East*, edited by Steven Heydemann, 100–48. Berkeley: University of California Press, 2000.

Vlaardingerbroek, Barend, and Yasmine Hachem-El-Masri. 'The Status of Evolutionary Theory in Undergraduate Biology Programs at Lebanese Universities: A Comparative Study'. *International Journal of Educational Reform* 15, No. 2 (2006): 150–63.

Vleuten, Erik van der. 'Toward a Transnational History of Technology: Meanings, Promises, Pitfalls'. *Technology and Culture* 49, No. 4 (2008): 974–94.

Waardenburg, Jean-Jacques. *Les universités dans le monde arabe actuel: Documentation et essai d'interprétation*. Vol. 1. Paris: Mouton, 1966.

Waast, Roland, and Pier-Luigi Rossi. 'Scientific Production in Arab Countries: A Bibliometric Perspective'. *Science Technology & Society* 15, No. 2 (2010): 339–70.

Wagner, Caroline, and Loet Leydesdorff. 'Network structure, self-organization, and the growth of international collaboration in science'. *Research Policy* 34, No. 10 (2005): 1608–18.

————. 'Mapping the network of global science: comparing international co-authorships from 1990 to 2000'. *International Journal of Technology and Globalisation* 1, No. 2 (2005): 185–208.

Wang, Zuoyue. 'Transnational Science during the Cold War: The Case of Chinese/American Scientists'. *Isis* 101, No. 2 (2010): 367–77.

Wazārat al-Maʿārif. *Ahamm miʾat ḥadath fī miʾat ʿām 1319-1419h*. Al-Riyāḍ: Rawnāʾ, 2000.

Willoughby, John. *Let a Thousand Models Bloom: Forging Alliances with Western Universities and the Making of the New Higher Educational System in the Gulf*. Washington: American University, 2008.

Zahlan, A.B. 'Planning science in the Arab world'. *Nature* 283 (1980): 239–41.

————. *Science and Science Policy in the Arab World*. London: Croom Helm, 1980.

Ziadat, Adel. *Western Science in the Arab World: The Impact of Darwinism, 1860–1930*. Basingstoke: Macmillan, 1986.

————.'Early Reception of Einstein's Relativity in the Arab Periodical Press'. *Annals of Science* 51, No. 1 (1994): 17–35.

# INDEX